ELECTRONICS EQUATIONS
HANDBOOK

ELECTRONICS EQUATIONS HANDBOOK

Stephen J. Erst

TAB Books
Division of McGraw-Hill, Inc.

New York   San Francisco   Washington, D.C.   Auckland   Bogotá
Caracas   Lisbon   London   Madrid   Mexico City   Milan
Montreal   New Delhi   San Juan   Singapore
Sydney   Tokyo   Toronto

JB

© 1989 by **TAB Books**.
TAB Books is a division of McGraw-Hill, Inc.

pbk   9 10 11 12 13 14 15 FGR/FGR 9 9 8 7 6
hc   1 2 3   4   5   6   7   8   9 FGR/FGR 8 9

Library of Congress Cataloging-in-Publication Data

Erst, Stephen J.
    Electronics equations handbook / by Stephen J. Erst.
        p.   cm.
    Bibliography: p. 235
    Includes index.
    ISBN 0-8306-9241-X   ISBN 0-8306-3241-7 (pbk.)
    1. Electronics—Mathematics.   I. Title.
    TK7835.E77   1989
    621.381′0151—dc20   89-5088
                                               CIP

Acquisitions Editor: Roland S. Phelps
Technical Editor and Book Design: Lisa A. Doyle
Director of Production: Katherine G. Brown

contents

acknowledgments

Equations (1-46) and (1-47) are reprinted with permission of *Electronic Design* vol. 12 no. 4 (February 17, 1964), VNU Business Publications, Inc. Equation (3-41) is courtesy *Ham Radio* (July 1981). Equations (13-44) and (13-45) are courtesy of D. E. Norton, "High Dynamic Range Transistor Amplifier Using Lossless Feedback," *Microwave Journal* (May 1976): 53–57.

introduction

The arena of electronics has grown at such a pace that those in practice need to continually refresh themselves with new information and store it in memory files or a library. Such storage presents the problem of recall, often with much frustration and lost time. This handbook was created to be a compilation of the most-used formulas and other information in a single volume to offer the user a first place to look, negating time-consuming searches through libraries and files.

This book begins with the relationships of the three simplest passive components—resistors, capacitors, and inductors—and progresses to more complex networks such as those involved with transmission lines, amplifiers, and modulation. The 21 chapters herewith include over 880 equations.

1

resistors, inductors, and capacitors

RESISTANCE (R)

RESISTIVITY

$$R = \rho l/a \tag{1-1}$$

where
ρ is the specific resistance or resistivity of the material
l is length of the conductor
a is the cross-sectional area of the conductor

The resistivity of a material is given in microhms ($\mu\Omega$) per centimeter cube. Copper has a resistivity of 1.589 microhms per centimeter cube at a temperature of 0°C and 1.724 microhms at 20°C.

TEMPERATURE DEPENDENCE OF RESISTIVITY

$$R_{t_1} = R_t[1 + \alpha_t(t_1 - t)] \tag{1-2}$$

where
R_{t_1} is the resistivity at temperature t_1
R_t is the resistivity at the reference temperature
α_t is the temperature coefficient at the reference temperature
t is the reference temperature

1

CONDUCTOR CROSS-SECTIONAL AREA

Conductor area is expressed in various forms depending on the shape factor of the particular case. For round wires, circular miles are used, and square mils are used for square conductors. The square mils notation is also used for rectangular cases.

Circular mils = 1.2732 square mils
Square mils = 0.7854 circular mils

RESISTANCES IN SERIES

$$R_{total} = R_1 + R_2 + R_3 + \cdots R_n \tag{1-3}$$

in *ohms*, where R_n represents the individual resistances, and n is the identifying resistance number.

RESISTANCES IN PARALLEL

$$R_{total} = \cfrac{1}{\cfrac{1}{R_1} + \cfrac{1}{R_2} + \cfrac{1}{R_3} + \cdots \cfrac{1}{R_n}} \tag{1-4}$$

in *ohms*, where R_n represents the individual resistances and n is the identifying number.

CONDUCTANCE IN PARALLEL

$$G_{total} = G_1 + G_2 + G_3 + \cdots G_n \tag{1-5}$$

G is in *siemens* (the old term used was *mhos*).

CONDUCTANCE (G)

Conductance (G) is the reciprocal of resistance. The unit is *siemens* (or *mhos*).

$$G = 1/R \tag{1-6}$$

DIRECT CURRENT RELATIONSHIPS

Ohms Law

$$E = IR \tag{1-7}$$

E = voltage (volts)
I = current (amperes)
R = resistance (ohms)

Alternate forms are

$$I = E/R \tag{1-8}$$
$$R = E/I \tag{1-9}$$

Conductance

$$G = 1/R \tag{1-10}$$

in *seimens* (or *mhos*).

Joule's Law

Power or rate of work in *watts* (W):

$$P = EI \tag{1-11}$$

$$P = EI \cdot 10^7 \text{ ergs per second} \tag{1-12}$$

The power converted into heat in t seconds where E and I are constant is

$$W = Pt \tag{1-13}$$

$$= EIt \cdot 10^7 \quad (\text{in } ergs) \tag{1-14}$$

$$= EIt \quad (\text{in } joules) \tag{1-15}$$

$$= I^2Rt \quad (\text{in } joules) \tag{1-16}$$

$$Q = I^2Rt/4.18 \quad (\text{in } gram\ calories) \tag{1-17}$$

$$= 0.24\ I^2R \quad (\text{in } gram\ calories) \tag{1-18}$$

Kirchhoff's Laws

First Law: *At any point in a circuit, the current flowing toward that point is equal to the current flowing away from the point.*

Second Law: *In any closed path or loop in a circuit, the sum of the IR drops must equal the sum of the applied voltages.*

AC POWER IN RESISTIVE CIRCUITS (SINUSOIDAL CASE)

$$P_{\text{peak}} = E_{\text{max}}I_{\text{max}} \tag{1-19}$$

$$P_{\text{average}} = \tfrac{1}{2}(E_{\text{max}}I_{\text{max}}) \tag{1-20}$$

$$= \frac{E_{\text{max}}}{\sqrt{2}} \cdot \frac{I_{\text{max}}}{\sqrt{2}} \tag{1-21}$$

$$= \text{rms voltage} \times \text{rms current} \tag{1-22}$$

The quantity rms is *root mean square* or *effective value*.

Form Factor (F_f)

$$F_f = \frac{\text{positive half cycle rms value}}{\text{positive half cycle average value}} \tag{1-23}$$

See Tables 1-1 and 1-2.

Table 1-1.   Voltage Conversion Factors for
Three Waveforms

Wave-form	To convert		
	rms to average	average to peak	rms to peak
	Multiply by		
Sine	$\dfrac{\pi}{2\sqrt{2}}$	$\dfrac{2}{\pi}$	$\dfrac{1}{\sqrt{2}}$
Triangle	$\dfrac{2}{\sqrt{3}}$	0.5	$\dfrac{1}{\sqrt{3}}$
Square	1	1	1

Table 1-2.   Decimal Conversion Factors for Sinusoidal Waveforms

From	(Multiply by)			
	To: rms	Average	Peak	Peak-to-peak
rms	1.0	0.9	1.414	2.828
Average	1.11	1.0	1.57	3.14
Peak	0.707	0.637	1.0	2.0
Peak-to-peak	0.3535	0.3185	0.5	1.0

INDUCTANCE (*L*)

INDUCTANCES IN SERIES

$$L_{\text{total}} = L_1 + L_2 + L_3 + \cdots L_n \quad \text{(in } henries) \tag{1-24}$$

INDUCTANCE IN PARALLEL

$$L_{\text{total}} = \frac{1}{\dfrac{1}{L_1} + \dfrac{1}{L_2} + \dfrac{1}{L_3} + \cdots \dfrac{1}{L_n}} \quad \text{(in } henries) \tag{1-25}$$

TWO INDUCTORS IN PARALLEL

$$L_{\text{total}} = \frac{L_1 L_2}{L_1 + L_2} \quad \text{(in } henries) \tag{1-26}$$

ALTERNATING CURRENT EXCITATION

Current and Voltage: Inductance Case

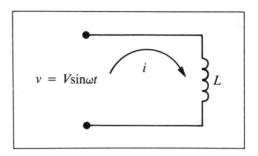

v is the instaneous voltage (in *volts*)
i is the instantaneous current (in *amperes*)
L is the inductance (in *henries*)

$$i = \frac{V}{X_l} \sin(\omega t - 90°) \tag{1-27}$$

$$i = I \sin(\omega t - 90°) \tag{1-28}$$

$$X_l = 2\pi f L \tag{1-29}$$

I is the peak current
V is the peak voltage
f is the frequency (in *hertz*)

$$v = L\frac{di}{dt} \tag{1-30}$$

Impedance

$$Z_l = \omega L \underline{/90°} \tag{1-31}$$

Instantaneous Power

$$p = vi \tag{1-32}$$

$$p = -\tfrac{1}{2}VI \sin(2\omega t) \tag{1-33}$$

Average Power

$$p_{\text{avg}} = 0 \tag{1-34}$$

Energy

$$W = \int_a^b p\,dt \tag{1-35}$$

where a and b define the time slot of the integration and p is the instantaneous power.

For a positive half cycle of power:

$$W = \tfrac{1}{2}L\,I^2 \qquad \text{(in } \textit{joules}\text{)}$$

L is in *henries*
I is in *amperes*

INDUCTORS AT RADIO FREQUENCIES

Quality Factor (Q_0)

$$Q_o = \frac{\omega L_o}{R_o} \tag{1-37}$$

$\omega = 2\pi f_0$
L_o is inductance (in *henries*)
R_o is the resistance of the inductor (in *ohms*)
f_o is frequency (in *hertz*)

Distributed Capacitance (C_d)

$$C_d = \frac{\pi D}{3.6 \cosh^{-1}(s/d)} \qquad \text{(in pF)} \tag{1-38}$$

D is the diameter of the coil at the wire center (in *mm*)
s is the spacing between turns at the wire centers (in *mm*)
d is the wire size (in *mm*)
$\cosh^{-1}$ is the inverse hyperbolic cosine

Self-Resonant Frequency (f_o)

$$f_o = \frac{1}{2\pi\sqrt{L_oC_d}} \qquad \text{(in } \textit{hertz}\text{)} \tag{1-39}$$

Apparent Inductance and Q

Condition: $f < 0.8\,f_o$

$$L = \frac{L_o}{1 - f/f_o} \tag{1-40}$$

$$Q = Q_o[1 - (f/f_o)^2]$$

Inductance Design

Single-Layer Coil

$$L = \frac{a^2n^2}{9a + 10b} \tag{1-41}$$

L is inductance (in *microhenries*)
a is the coil radius (in *inches*)
b is the coil length (in *inches*)
n is the number of turns

In alternate form:

$$n = \frac{[L(9a + 10b)]^{\frac{1}{2}}}{a}$$ (1-42)

Straight Wire in Free Space

$$L = 2 \cdot 10^{-4} \, l \left[\left(\ln 2\frac{l}{r} \right) - \frac{3}{4} \right]$$ (1-43)

or

$$L = 2 \cdot 10^{-4} \, l \left[\left(2.303 \log_{10} 2\frac{l}{r} \right) - \frac{3}{4} \right]$$ (1-44)

r is the wire radius (in *mm*)
l is the wire length (in *mm*)

Straight Wire Parallel to a Ground Plane with One End Grounded

$$L = 4605 \cdot 10^{-7} \, l \left\{ \log_{10} \left[\frac{2h}{r} \left(\frac{l + k}{l + p} \right) \right] \right\} +$$
$$2 \cdot 10^{-4} \, (p - k + 0.25l - 2h + r)$$ (1-45)

where
$k = (l^2 + r^2)^{\frac{1}{2}}$
$p = (l^2 - 4h^2)^{\frac{1}{2}}$
h is the wire height above ground (in *mm*)
l is the wire length (in *mm*)
r is the wire radius (in *mm*)

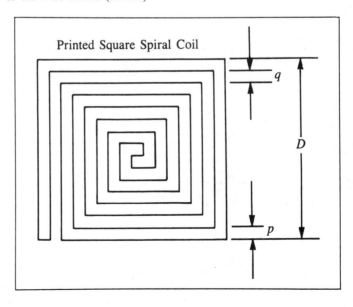

Printed Square Spiral Coil

Inductance (in henries)

$$L = 85 \cdot 10^{-10} \, DN^{\frac{3}{3}} \tag{1-46}$$

or

$$L = 27 \cdot 10^{-10} \, (D^{\frac{8}{3}}/p^{\frac{3}{3}})(1 + R^{-1})^{\frac{3}{3}} \tag{1-47}$$

D is dimension of the square coil (in *cm*)
N is the number of turns.
$R = p/q$
$p = q$

MUTUAL INDUCTANCE

The mutual inductance between two inductors is:

$$M = k\sqrt{L_1 L_2} \tag{1-48}$$

where
M is the mutual inductance
k is the coefficient of coupling (maximum coupling $k = 1$)
L_1 and L_2 are the inductance of each inductor.

Inductors in Series with Mutual Inductance

Additive:

$$L_{\text{total}} = L_1 + L_2 + 2M \tag{1-49}$$

Subtractive:

$$L_{\text{total}} = L_1 + L_2 - 2M$$

CAPACITANCE (C)

UNITS

farads (F)
microfarads (μF) = 10^{-6}
picofarads (pF) = 10^{-12}

SERIES CAPACITORS

$$C_{\text{total}} = \cfrac{1}{\cfrac{1}{C_1} + \cfrac{1}{C_2} + \cfrac{1}{C_3} + \cdots \cfrac{1}{C_n}} \tag{1-50}$$

PARALLEL CAPACITORS

$$C_{\text{total}} = C_1 + C_2 + C_3 + \cdots C_n \tag{1-51}$$

Two Capacitors in Series

$$C_{total} = \frac{C_1 C_2}{C_1 + C_2} \tag{1-52}$$

Charge of a Capacitor (Q)

$$Q = CE \quad \text{(in } coulombs\text{)} \tag{1-53}$$

C is capacitance (in *farads*)
E is the applied potential (in *volts*)

Energy Stored by a Capacitor (W)

$$W = \tfrac{1}{2}(CE^2) \quad \text{(in } joules \text{ or } watt\text{-}seconds\text{)} \tag{1-54}$$
$$= Q^2/(2C) \tag{1-55}$$
$$= \tfrac{1}{2}QE \tag{1-56}$$

Alternating Current Excitation

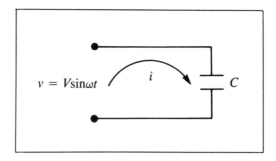

Instantaneous Current

$$i = I_{max} \sin (\omega t + 90°) \tag{1-57}$$

Instantaneous Voltage Across C

$$v = V \sin \omega t \tag{1-58}$$

Impedance

$$Z_c = 1/(\omega C)\angle{-90°} \tag{1-59}$$

Reactance

$$X_C = 1/(\omega C) \tag{1-60}$$

Instantaneous Power

$$P = vi \tag{1-61}$$
$$P = \tfrac{1}{2}VI_{max} \sin 2\omega t \tag{1-62}$$

Energy

$$W_c = \int_{t_1}^{t_2} \frac{1}{2} EI_{max} \sin 2\omega t \; (dt) \tag{1-63}$$

The terms t_1 and t_2 define that portion of the instantaneous power to be integrated to determine the energy content in that time space.

CAPACITANCE OF PARALLEL PLATES

$$C = 8.84173 \cdot 10^{-3} \frac{KA}{g} \quad (\text{in } pF) \tag{1-64}$$

A is area (in cm^2)
K is dielectric constant
g is the separation between plates (in cm)
$A^{\frac{1}{2}} > g$

Alternate Form:

$$C = 0.2247 \frac{KA}{g} \quad (\text{in } pF)$$

g and A are in *inches*

CAPACITANCE OF COAXIAL CABLES OR CYLINDERS

$$C = 7.354 \frac{K}{\log_{10} \dfrac{D}{d}} \quad (\text{in } pF) \tag{1-65}$$

D is the inner diameter of the outer cylinder (in *inches*)
d is the outer diameter of the inner cylinder (in *inches*)

REACTANCE (X)

CAPACITIVE

$$X_C = \frac{1}{2\pi fC} \quad (\text{in } ohms) \tag{1-66}$$

f is frequency
C is capacitance (in *farads*)

The sign of capacitive reactance is negative by convention.

INDUCTIVE

$$X_L = 2\pi fL \quad (\text{in } ohms) \tag{1-67}$$

f is frequency (in *Hz*)

L is inductance (in *henries*)

REACTANCE IN SERIES OF THE SAME KIND

$$X_T = X_1 + X_2 + X_3 + \cdots X_n \tag{1-68}$$

REACTANCE IN PARALLEL OF THE SAME KIND

$$X_T = \cfrac{1}{\cfrac{1}{X_1} + \cfrac{1}{X_2} + \cfrac{1}{X_3} + \cdots \cfrac{1}{X_4}} \tag{1-69}$$

TWO REACTANCES IN PARALLEL OF THE SAME KIND

$$X_T = \frac{X_1 X_2}{X_1 + X_2} \tag{1-70}$$

REACTANCES OF DIFFERENT KINDS IN SERIES

$$X_T = X_L - X_C \tag{1-71}$$

REACTANCES OF DIFFERENT KINDS IN PARALLEL

$$X_T = \frac{-X_C X_L}{-X_C + X_L} \tag{1-72}$$

COMPLEX ALGEBRA

ADDITION

$$(R_1 + jX_1) + (R_2 + jX_2) = (R_1 + R_2) + j(X_1 + X_2) \tag{1-73}$$

$$j = \sqrt{-1}$$

SUBTRACTION

$$(R_1 + jX_1) - (R_2 + jX_2) = (R_1 - R_2) + j(X_1 - X_2) \tag{1-74}$$

MULTIPLICATION (POLAR FORM)

To convert to polar form, use

$$|Z| = \sqrt{R^2 + X^2} \tag{1-75}$$

and

$$\theta = \tan^{-1} \frac{X}{R} \tag{1-76}$$

Then,

$$(|Z_1| \angle \theta_1)(|Z_2| \angle \theta_2) = |Z_1||Z_2| \angle (\theta_1 + \theta_2) \tag{1-77}$$

DIVISION (POLAR FORM)

Convert to polar form as in equations (1-75) through (1-77). Then,

$$\frac{|Z_1|\angle\theta_1}{|Z_2|\angle\theta_2} = \frac{|Z_1|}{|Z_2|}\angle(\theta_1 - \theta_2) \tag{1-78}$$

MULTIPLICATION

$$(R_1 + jX_1)(R_2 + jX_2) = R_1R_2 + jR_1X_2 + jR_2X_1 - X_1X_2$$
$$= R_1R_2 - X_1X_2 + j(R_1X_2 + R_2X_1) \tag{1-79}$$

DIVISION

$$\frac{R_1 + jX_1}{R_2 + jX_2} = \frac{(R_1 + jX_1)}{(R_2 + jX_2)}\frac{(R_2 - jX_2)}{(R_2 - jX_2)}$$
$$= \frac{R_1R_2 - jR_1X_2 + jR_2X_1 + X_1X_2}{R_2{}^2 + X_2{}^2} \tag{1-80}$$

NOTATION CONVERSION

$$Z\angle\theta = \sqrt{R^2 + X^2}\angle\tan^{-1}\frac{X}{R}$$
$$= Z\cos\theta + jZ\sin\theta \tag{1-81}$$

J OPERATOR

The term j is used to define the quadrant location of a vector and is defined as:

$$j^2 = -1$$

j defines a positive 90° vector rotation
$-j$ defines a negative 90° vector rotation

VECTOR COORDINATES

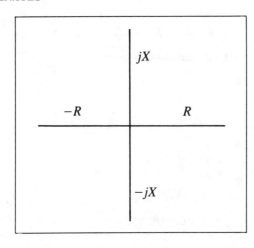

ALTERNATING CURRENT IN IMPEDANCE

Z = Impedance
E = Voltage
I = Current
θ = Phase angle

$E = IZ$	$E = P/I \cos \theta$	$E = \sqrt{PZ}/\cos \theta$	(1-82)
$I = E/Z$	$I = P/E \cos \theta$	$I = \sqrt{P}/\cos \theta$	(1-83)
$Z = E/I$	$Z = (E^2 \cos \theta)/P$	$Z = P/I^2 \cos \theta$	(1-84)
$P = IE \cos \theta$	$P = (E^2 \cos \theta)/Z$	$P = I^2Z \cos \theta$	(1-85)

RC CIRCUITS

CHARGE OF AN RC CIRCUIT

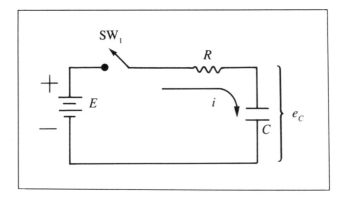

Conditions:

C is discharged
SW_1 is open at $t = 0$

When SW_1 is closed,

$$i = \frac{E}{R}e^{-t/(RC)} \tag{1-86}$$

$e = 2.718$
t is any time after SW_1 is closed
R is resistance (in *ohms*)
C is capacitance (in *farads*)
E is the applied voltage (in *volts*)

$$e_C = E - iR \tag{1-87}$$
$$= E(1 - e^{-t/(RC)}) \tag{1-88}$$

The final charge Q of capacitor C is equal to CE. The instantaneous charge q is

$$q = CE(1 - e^{-t/(RC)}) \tag{1-89}$$

or

$$q = Q(1 - e^{-t/(RC)}) \tag{1-90}$$

RC CIRCUIT DISCHARGE

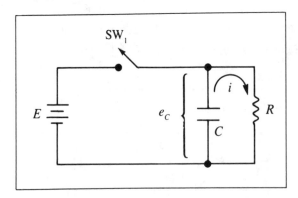

Conditions:
    C is fully charged
    SW_1 is closed at $t = 0$

When SW_1 is opened,

$$i = \frac{E}{R}e^{-t/(RC)} \tag{1-91}$$

and

$$e_C = iR \tag{1-92}$$
$$= Ee^{-t/(RC)} \tag{1-93}$$

The instantaneous charge on C at any time t is

$$q = C\,e_C \tag{1-94}$$
$$= Q\,e^{-t/(RC)} \quad \text{(since } Q = CE\text{)} \tag{1-95}$$

TIME CONSTANT (T)

The time constant T is that value of time required for an RL or RC circuit to reach a voltage or current level equal to 63.2 percent of its steady-state value when a DC voltage is applied. Alternately, when the DC voltage is removed, T is that time when the voltage or current level falls to 36.8 percent of the steady-state value.

For RC and RL circuits, a value of T equal to or greater than five time constants is essentially 100 percent of the final value. For charging circuitry, the final value is the full value of the applied voltage, while for the discharge configuration, the final value is zero.

RC COUPLING PULSE DROOP (GENERAL CASE)

The coupling droop of a pulse can be found from

$$e_o(t) = e_s \frac{R_l}{R_s + R_l} \epsilon^k \qquad (1\text{-}96)$$

$$k = -\frac{T}{C(R_s + R_l)} \qquad (1\text{-}97)$$

$$0 \le t \le T$$

where

e_o is the output of the RC coupling circuit
R_l is the load resistance
R_s is the source resistance
C is the coupling capacitance
T is the pulse width
ϵ is the base of natural logarithms (2.71828183)
k is the exponent of ϵ of equation (1-96) and is defined in
equation (1-97)

RC COUPLING PULSE DROOP (10 PERCENT)

For a pulse trailing edge droop of 10 percent, the values of R and C are related by

$$C(R_s + R_l) \cong 10T \qquad (1\text{-}98)$$

where

C is the coupling capacitance
R_s is the source resistance
R_l is the load resistance
T is the pulse width

PULSE DROOP

This equation determines the capacitance between two impedances to result in no more than 10 percent pulse droop.

T is pulse width

$$C = \frac{10T}{R_s + R_l} \qquad (1\text{-}99)$$

EXAMPLE:

$$T = 100 \ \mu s$$
$$R_s = R_l = 50 \text{ ohms}$$
$$C = \frac{1000 \cdot 10^{-6}}{100} = 10 \ \mu F \qquad (1\text{-}100)$$

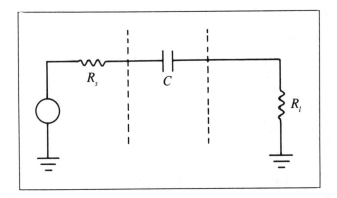

PULSE RISE TIME (BAND LIMITED CASE)   (t_r)

$t_r = 0.35/f_{max}$ (1-101)

f_{max} is the upper -3dB point of the frequency response of the network or amplifier

PULSE TILT (BAND LIMITED CASE)

$\text{Tilt} = 628.32 \, f_{low}/T$ (1-102)

f_{low} is the lower -3dB point of the network or amplifier
T is the pulse width

RL CIRCUITS _____

RL SERIES CIRCUITS

Voltage and Current

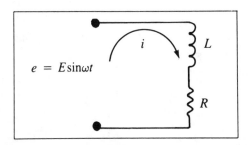

$e = E \sin \omega t$

$\quad = L\dfrac{di}{dt} + R \cdot i$

$\quad = \omega L I \cos \omega t + R I \sin \omega t$ (1-103)

where $i = I \sin \omega t$

Phase Angle

$$\theta = \tan^{-1}\left(\frac{\omega L}{R}\right)$$

(1-104)

Impedance

$$Z_{LR} = \sqrt{R^2 + (\omega L)^2}\ \underline{\left|\tan^{-1}\left(\frac{\omega L}{R}\right)\right.}$$

(1-105)

Instantaneous Power

$$p = ei$$

(1-106)

$$= \frac{EI}{2}\ [\cos\theta - (\cos 2\omega t)\cos\theta + (\sin 2\omega t)\sin\theta]$$

Average Power

$$P_{\text{avg}} = \frac{EI}{2}\cos\theta$$

(1-107)

Volt-Amperes (*VA*)

$$\text{Real Power:}\ \frac{EI}{2}\cos\theta$$

(1-108)

$$\text{Reactive Power:}\ \frac{EI}{2}\sin\theta$$

(1-109)

$$VA = \sqrt{\left(\frac{EI}{2}\cos\theta\right)^2 + \left(\frac{EI}{2}\sin\theta\right)^2}$$

(1-110)

$$= \frac{EI}{2}$$

(1-111)

Instantaneous Power (Real)

$$p_{\text{real}} = \frac{EI}{2}\cos\theta\ (1 - \cos 2\omega t)$$

(1-112)

Instantaneous Power (Reactive)

$$p_{\text{reactive}} = \frac{EI}{2}\sin\theta\ (\sin 2\omega t)$$

(1-113)

RESONANT CIRCUITS

SERIES

R is the resistance of the inductor alone or in combination with an additional resistance external to the inductor.

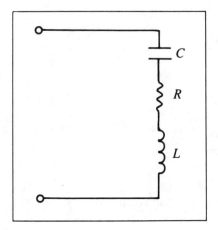

Resonant Frequency

$$f = \frac{1}{2\pi(LC)^{\frac{1}{2}}}$$

(1-114)

Impedance at Any Frequency

$$Z = \sqrt{R^2 + (X_L - X_C)^2}$$

(1-115)

where

$$X_L = 2\pi f L$$

(1-116)

$$X_C = \frac{1}{2\pi f C}$$

(1-117)

At resonance, $Z = R$.

Sharpness of Resonance (Q)

$$Q = X_L/R$$

(1-118)

$$Q = \frac{2\pi f L}{R}$$

(1-119)

The larger Q is, the sharper the circuit resonance will be.

PARALLEL RESONANCE

Resonant Frequency

$$f = \frac{1}{2\pi(LC)^{\frac{1}{2}}}$$

(1-120)

Impedance at Resonance

$$Z = \frac{(X_L)^2}{R}$$

(1-121)

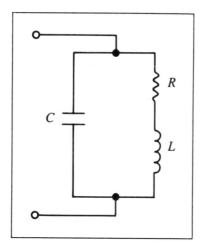

$$Z = \frac{(2\pi fL)^2}{R} \qquad (1\text{-}122)$$

or

$$Z = QX_L \qquad (1\text{-}123)$$

$$Z = Q\, 2\pi fL \qquad (1\text{-}124)$$

Parallel Resonant RLC Network

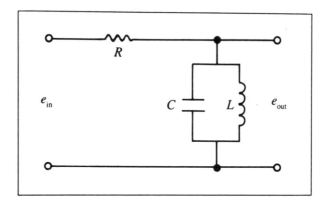

Transfer Function

$$F(S) = \frac{\dfrac{SL}{R}}{S^2LC + \dfrac{SL}{R} + 1} \qquad (1\text{-}125)$$

Frequency Response

Magnitude

$$|F(j\omega)| = \frac{\dfrac{\omega L}{R}}{\left[(1 - \omega^2 LC)^2 + \dfrac{\omega^2 L^2}{R^2} \right]^{\frac{1}{2}}} \tag{1-126}$$

Phase

$$\theta(\omega) = \frac{\pi}{2} - \tan^{-1}\left(\frac{\dfrac{\omega L}{R}}{1 - \omega^2 LC} \right) \tag{1-127}$$

2

---

networks

RESISTIVE ATTENUATORS

Pi Attenuator

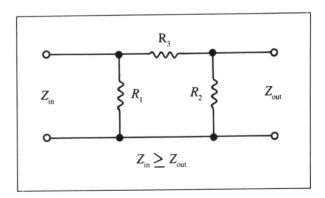

$$1/R_1 = (1/Z_{in})k - 1/R_3 \tag{2-1}$$

$$1/R_2 = (1/Z_{out})k - 1/R_3 \tag{2-2}$$

$$R_3 = \frac{1}{2}(L - 1)\left(\frac{Z_{in}Z_{out}}{L}\right)^{\frac{1}{2}} \tag{2-3}$$

L is the desired loss ratio

The minimum loss is:

$$L_{min} = 10 \log_{10} \left[\left(\frac{Z_{in}}{Z_{out}} \right)^{\frac{1}{2}} + \left(\frac{Z_{in}}{Z_{out}} - 1 \right)^{\frac{1}{2}} \right]^2 \quad \text{(in } dB\text{)} \tag{2-4}$$

$$k = \frac{L+1}{L-1}$$

$$L_{ratio} = \text{anl} \frac{L_{dB}}{10}$$

Anl (or $\log^{-1}$) is the antilogarithm or antilog

T ATTENUATOR

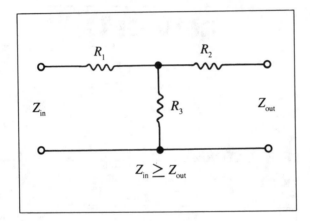

$$R_1 = Z_{in} \, k - R_3 \tag{2-5}$$
$$R_2 = Z_{out} \, k - R_3 \tag{2-6}$$
$$R_3 = \frac{2(L \, Z_{in} Z_{out})^{\frac{1}{2}}}{L-1} \tag{2-7}$$
$$k = \frac{L+1}{L-1} \tag{2-8}$$

L is the desired loss (ratio)

The minimum loss is:

$$L_{min} = 10 \log_{10} \left[\left(\frac{Z_{in}}{Z_{out}} \right)^{\frac{1}{2}} + \left(\frac{Z_{in}}{Z_{out}} - 1 \right)^{\frac{1}{2}} \right]^2 \quad \text{(in } dB\text{)} \tag{2-9}$$

$$L_{ratio} = \text{anl} \frac{L_{dB}}{10}$$

anl is the inverse logarithm (or $\log^{-1}$)

AC VOLTAGE DIVIDERS

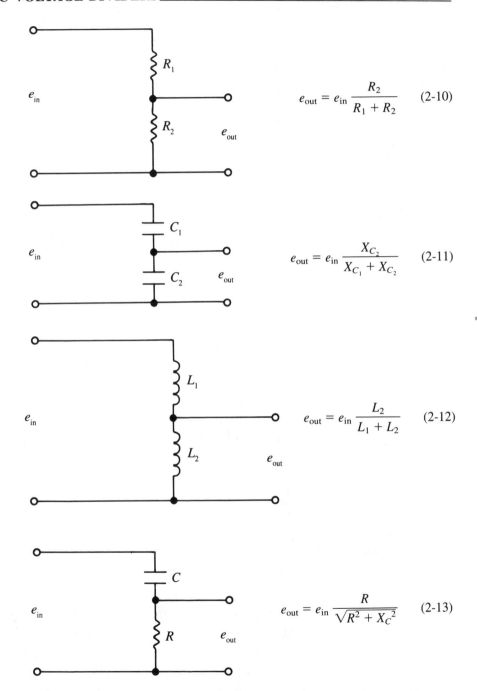

$$e_{\text{out}} = e_{\text{in}} \frac{R_2}{R_1 + R_2} \qquad (2\text{-}10)$$

$$e_{\text{out}} = e_{\text{in}} \frac{X_{C_2}}{X_{C_1} + X_{C_2}} \qquad (2\text{-}11)$$

$$e_{\text{out}} = e_{\text{in}} \frac{L_2}{L_1 + L_2} \qquad (2\text{-}12)$$

$$e_{\text{out}} = e_{\text{in}} \frac{R}{\sqrt{R^2 + X_C^{\,2}}} \qquad (2\text{-}13)$$

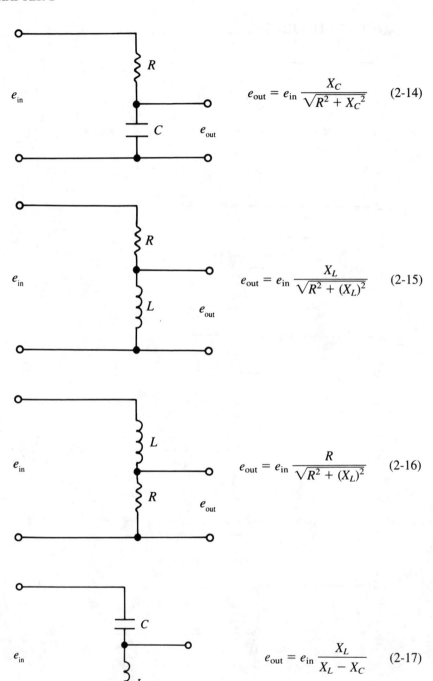

$$e_{\text{out}} = e_{\text{in}} \frac{X_C}{\sqrt{R^2 + X_C{}^2}} \qquad (2\text{-}14)$$

$$e_{\text{out}} = e_{\text{in}} \frac{X_L}{\sqrt{R^2 + (X_L)^2}} \qquad (2\text{-}15)$$

$$e_{\text{out}} = e_{\text{in}} \frac{R}{\sqrt{R^2 + (X_L)^2}} \qquad (2\text{-}16)$$

$$e_{\text{out}} = e_{\text{in}} \frac{X_L}{X_L - X_C} \qquad (2\text{-}17)$$

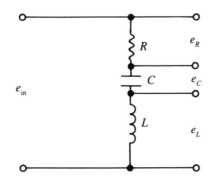

$$e_L = e_{in} \frac{X_L}{R^2 + (X_L - X_C)^2} \qquad (2\text{-}18)$$

$$e_c = e_{in} \frac{X_C}{\sqrt{R^2 + (X_L - X_C)^2}} \qquad (2\text{-}19)$$

$$e_R = e_{in} \frac{R}{\sqrt{R^2 + (X_L - X_C)^2}} \qquad (2\text{-}20)$$

$$e_C + e_L = e_{in} \frac{X_L - X_C}{\sqrt{R^2 + (X_L - X_C)^2}} \qquad (2\text{-}21)$$

T TO PI AND PI TO T TRANSFORMATION_____

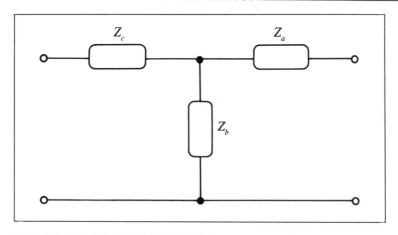

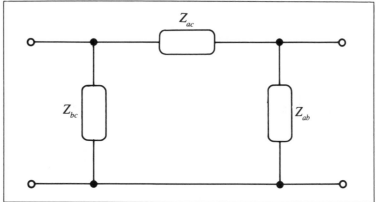

T to Pi

$$Z_{ab} = k/Z_c \qquad (2\text{-}22)$$

$$Z_{bc} = k/Z_a \qquad (2\text{-}23)$$

$$Z_{ac} = k/Z_b \tag{2-24}$$

$$k = Z_aZ_b + Z_bZ_c + Z_aZ_c \tag{2-25}$$

Pi to T

$$Z_a = Z_{ab}Z_{ac}/p \tag{2-26}$$

$$Z_b = Z_{ab}Z_{bc}/p \tag{2-27}$$

$$Z_c = Z_{ac}Z_{bc}/p \tag{2-28}$$

$$p = Z_{ab} + Z_{bc} + Z_{ac} \tag{2-29}$$

LADDER NETWORKS

Adding a series R, L, or C to a circuit or network with an impedance Z results in the following. The term Y is the admittance.

$$Y = \frac{1}{Z} \tag{2-30}$$

Series R

$$Y_1 = \left[\left(\frac{1}{Y} \right) + \left(R + j0 \right) \right]^{-1} \tag{2-31}$$

Series L

$$Y_1 = \left[\left(\frac{1}{Y} \right) + \left(0 + j\omega L \right) \right]^{-1} \tag{2-32}$$

Series C

$$Y_1 = \left[\left(\frac{1}{Y} \right) + \left(0 - j\frac{1}{\omega C} \right) \right]^{-1} \tag{2-33}$$

Parallel R

$$Y_1 = Y + \left(\frac{1}{R} + j0 \right) \tag{2-34}$$

Parallel L

$$Y_1 = Y + \left(0 - j\frac{1}{\omega L} \right) \tag{2-35}$$

Parallel C

$$Y_1 = Y + (0 + j\omega C) \tag{2-36}$$

The new $Z_1 = 1/Y_1$. $\tag{2-37}$

QUARTER-WAVE LINE-MATCHING SECTIONS

Two different impedances can be matched by placing a quarter-wave section of line with the impedance $Z = (Z_{out} \cdot Z_L)^{1/2}$ between them as shown below.

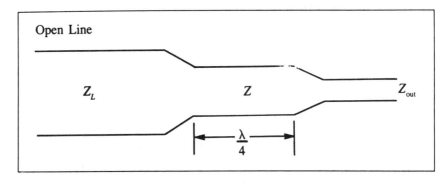

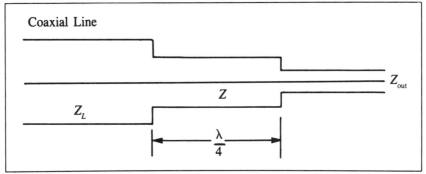

BROADBAND TERMINATION INTERSTAGE MATCH

This circuit provides a broadband 50-ohm match and is useful for mixer and preamplifier termination. An example is the termination of a mixer before presenting its output to a crystal filter.

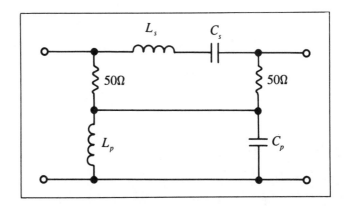

where

$$X_{L_s} = X_{C_s} = 50Q \tag{2-38}$$

$$X_{L_p} = X_{C_p} = 50/Q \tag{2-39}$$

s represents series elements

p represents parallel elements

Q is the quality factor

CAPACITANCE DIVISION IMPEDANCE MATCHING

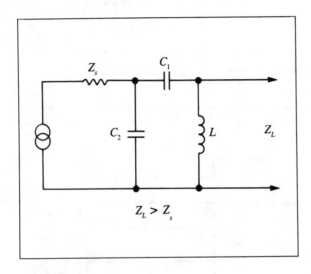

Approximate Relationship

$$Z_L = Z_s \left(\frac{C_1 + C_2}{C_1}\right)^2 \tag{2-40}$$

Alternate Approximation

Find C_t which resonates with L. Compute

$$P = \frac{\text{high } Z}{\text{low } Z} \tag{2-41}$$

Then

$$C_2 = P\,C_t \tag{2-42}$$

$$C_1 = \left(\frac{P}{P-1}\right)C_t \tag{2-43}$$

Exact Form

$$Z_L = \frac{(X_{C_1})^2}{Z_s} + Z_s\left(\frac{C_1 + C_2}{C_1}\right)^2 \tag{2-44}$$

PI NETWORK EQUATIONS

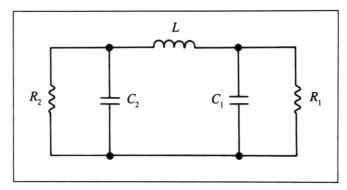

$$X_{C_2} = R_2/Q \qquad \qquad (2\text{-}45)$$

$$X_L = \frac{Q\,R_2 + R_1\,R_2/X_{C_1}}{Q^2 + 1} \qquad \qquad (2\text{-}46)$$

$$X_{C_1} = R_1 \sqrt{\frac{R_2}{R_1(Q^2 + 1) - R_2}} \qquad \qquad (2\text{-}47)$$

where $10 < Q < 20$

MATCHING NETWORKS

SOURCE REPRESENTATION (SERIES R_1, C_o)

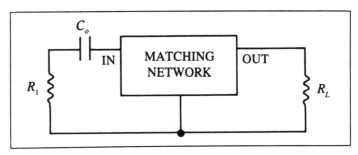

Circuit

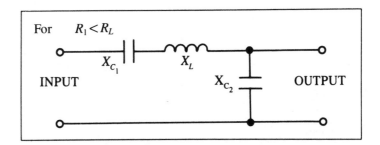

$$X_{C_1} = Q R_1 \quad \text{(select } Q\text{)} \tag{2-48}$$

$$X_L = X_{C_1} + \left(\frac{R_1 R_L}{X_{C_2}}\right) + X_{C_o} \tag{2-49}$$

$$X_{C_2} = R_L \left[\frac{R_1}{R_L - R_1}\right]^{\frac{1}{2}} \tag{2-50}$$

When $R_1 < R_L$

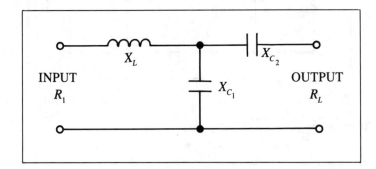

$$X_L = Q R_1 + X_{C_o} \quad \text{(select } Q\text{)} \tag{2-51}$$

$$X_{C_1} = \frac{B}{Q - A} \tag{2-52}$$

$$B = R_1(1 + Q^2) \tag{2-53}$$

$$A = \left\{\left[\frac{R_1(1 + Q^2)}{R_L}\right] - 1\right\}^{\frac{1}{2}} \tag{2-54}$$

$$X_{C_2} = A R_L \tag{2-55}$$

When $R_1 \neq R_L$

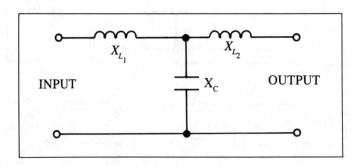

$$X_{L_1} = R_1 Q + X_{C_o} \quad \text{(select } Q\text{)} \tag{2-56}$$

$$X_C = \frac{A}{Q + B} \tag{2-57}$$

$$X_{L_1} = R_1 Q + X_{C_o} \qquad \text{(select } Q)$$ (2-58)

$$A = R_1(1 + Q^2)$$ (2-59)

$$B = \left[\frac{A}{R_L} - 1\right]^{\frac{1}{2}}$$ (2-60)

SOURCE REPRESENTATION FOR R_1 AND C_o PARALLEL

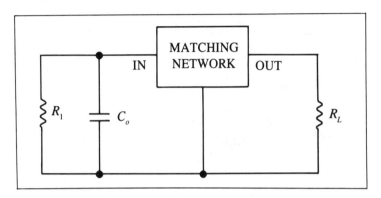

When $R_1 < R_L$

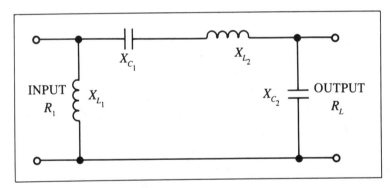

$$X_{L_1} = X_{C_o}$$

$$X_{C_1} = Q R_1 \qquad \text{(select } Q)$$ (2-61)

$$X_{L_2} = X_{C_1} + \left(\frac{R_1 R_L}{X_{C_2}}\right)$$ (2-62)

$$X_{C_2} = R_L \left(\frac{R_1}{R_L - R_1}\right)^{\frac{1}{2}}$$ (2-63)

When $R_1 > R_L$

$$X_{C_1} = \frac{R_1}{Q} \qquad \text{(select } Q)$$ (2-64)

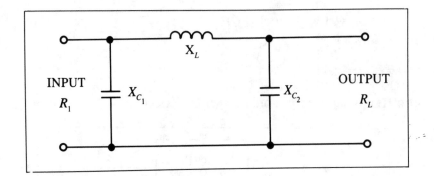

$$X_L = \frac{Q\,R_1 + (R_1 \cdot R_L)/X_{C_2}}{Q^2 + 1} \qquad\qquad (2\text{-}65)$$

$$X_{C_2} = R_L\left[\frac{R_1/R_L}{(Q^2 + 1) - (R_1/R_L)}\right]^{\frac{1}{2}} \qquad\qquad (2\text{-}66)$$

3

transmission lines

COAXIAL CABLE EQUATIONS

Variables are defined after equation (3-11).

IMPEDANCE

$$Z_o = \sqrt{\frac{L}{C}} = \frac{138}{\sqrt{E}} \log (D/d) \quad (\text{in } ohms) \tag{3-1}$$

CAPACITANCE

$$C = \frac{7.36 \, E}{\log (D/d)} \quad (\text{in } picofarads/ft) \tag{3-2}$$

INDUCTANCE

$$L = 0.14 \log (D/d) \quad (\text{in } microhenries/ft) \tag{3-3}$$

CUTOFF FREQUENCY

$$F_{co} = \frac{7.5}{\sqrt{E} \, (D + d)} \quad (\text{in } GHz) \tag{3-4}$$

TIME DELAY

$$T_d = 1.016 \sqrt{E} \quad (\text{in } nanoseconds/ft) \tag{3-5}$$

VELOCITY OF PROPAGATION (PERCENT OF THE SPEED OF LIGHT)

$$V_p = 100/\sqrt{E} \quad \text{(in } percent) \tag{3-6}$$

REFLECTION COEFFICIENT

$$\Gamma = \frac{Z_r - Z_o}{Z_r + Z_o} \tag{3-7}$$

$$\Gamma = \frac{\text{VSWR} - 1}{\text{VSWR} + 1} \tag{3-8}$$

VOLTAGE STANDING WAVE RATIO

$$\text{VSWR} = \frac{1 + \Gamma}{1 - \Gamma} \tag{3-9}$$

PEAK VOLTAGE ON THE CABLE

$$V_p = \frac{1.15 \, d(\log D/d)S}{K} \quad \text{(in } volts) \tag{3-10}$$

ATTENUATION PER 100 FEET

$$\alpha = \frac{0.435}{Z_o D} \left(\frac{D}{d} \cdot K_1 + K_2 \right) \sqrt{F} + 2.78\sqrt{E} \cdot P_f \cdot F \tag{3-11}$$

where

E is the cable insulation dielectric constant
D is the inside diameter of the outer conductor (in *inches*)
d is the outside diameter of the inner conductor (in *inches*)
Z_r is the terminating impedance
S is the breakdown voltage of the cable insulation (in *volts/mil*)
K is the safety factor
K_1 is the strand factor
K_2 is the braid factor
P_f is the power factor
F is the frequency (in *Hz*)

Table 3-1.
Dielectric Constant of
Insulating Materials

Insulation	Dielectric Constant (E)	Velocity (V_p)
TFE	2.1	69
ethylene propylene	2.24	66.8
polyethylene	2.3	65.9
cellular polyethylene	1.4-2.1	84.5-69.0
silicone rubber	2.08-3.5	69.3-53.4
polyvinylchloride	3-8	57.7-35.4

SLOTTED AND OPEN LINES OF IMPEDANCE Z_o

For $L < \lambda/4$

$$Z_s \quad = \quad jZ_o \tan \beta l \qquad (3\text{-}12)$$

$$Z_s \quad = \quad -jZ_o \cot \beta l \qquad (3\text{-}13)$$

For $l = \lambda/4$

$$Z_s \quad = \quad = \frac{2Z_o{}^2}{Rl} \quad \text{(approx.)} \qquad (3\text{-}14)$$

$$Z_s \quad = \quad = 0 \qquad (3\text{-}15)$$

For $(\lambda/2) > l > (\lambda/4)$

$$Z_s \quad = \quad = jZ_o \cot \beta l \qquad (3\text{-}16)$$

$$Z_s \quad = \quad = -j \tan \beta l \qquad (3\text{-}17)$$

For $l = \lambda/2$

$$Z_s \quad = \quad = \frac{2Z_o{}^2}{Rl} \text{(approx.)} \quad (3\text{-}18)$$

Note: A line that is $\lambda/2$ repeats the load.

$\beta = \omega\sqrt{LC}$
$\omega = 2\pi f$
l is length
R is loss per unit length

SINGLE-WIRE ABOVE-GROUND TRANSMISSION LINE_____

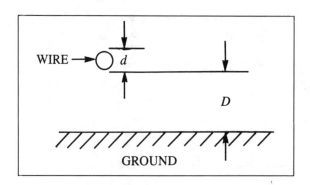

$$Z_o = 138 \log_{10} K \qquad (in \ ohms) \tag{3-19}$$

$$C = 24.12 \ \frac{1}{\log_{10}(K)} \qquad (in \ pF \ per \ meter) \tag{3-20}$$

$$L = 0.460 \log_{10}(K) \qquad (in \ \mu H \ per \ meter) \tag{3-21}$$

$$K = \frac{4D}{d} \tag{3-22}$$

$$p = 8.3 \ \frac{f^{\frac{1}{2}}}{d} \qquad (in \ \mu\Omega \ per \ meter) \tag{3-23}$$

= the resistivity of copper

Dimensions are in centimeters except as stated.

TWO-WIRE LINE_____

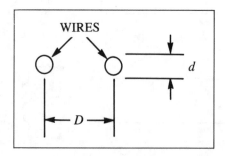

$$Z_o = 276 \log_{10} K \quad \text{(in } ohms) \tag{3-24}$$

$$C = 12.06 \frac{1}{\log_{10} K} \quad \text{(in } pF \text{ per meter}) \tag{3-25}$$

$$L = 0.92 \log_{10} K \quad \text{(in } \mu H \text{ per meter}) \tag{3-26}$$

$$K = \frac{2D}{d} \tag{3-27}$$

$$\rho = 16.6 \frac{f^{\frac{1}{2}}}{d} \quad \text{(in } \mu\Omega \text{ per meter for copper}) \tag{3-28}$$

The two-wire transmission line has less loss than a single wire above ground. Dimensions are in centimeters.

PARALLEL STRIP LINE IMPEDANCE

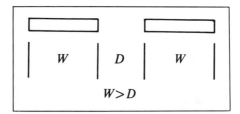

$$Z_o = 377 \frac{D}{W} \tag{3-29}$$

where

    D is the distance between the strips

    W is the width of the strips

DELAY LINES

DISTRIBUTED CONSTANT DELAY LINES

The construction is of a solonoidal shield over a slotted center conductor, separated by a dielectric. It's useful range of output impedance is 350 to 2000 ohms.

Equivalent Circuit

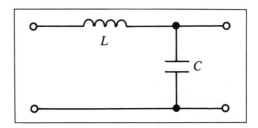

Time Delay (T_d)

$$T_d = (LC)^{\frac{1}{2}} \tag{3-30}$$

Impedance (Z_o)

$$Z_o = (L/C)^{\frac{1}{2}} \tag{3-31}$$

Low-Frequency Inductance (L)

$$L = 10^{-9}\pi^2 n^2\, d^2 K \quad \text{(in } henries\text{)} \tag{3-32}$$

d is the diameter of the coil
K is Nagaka's constant
n is the number of turns

Note: K approaches 1 as the diameter-to-length ratio approaches 0.

LUMPED CONSTANT DELAY LINES

Low-Pass Filter as a Delay Line

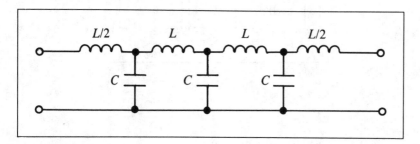

$$Z_o = (L/C)^{\frac{1}{2}}[1 - (\omega/\omega_c)]^{\frac{1}{2}} \tag{3-33}$$

Time Delay at ω

$$\tau = n(L/C)^{\frac{1}{2}} \left(\frac{1}{[1 - (\omega/\omega_c)^2]^{\frac{1}{2}}} \right) \tag{3-34}$$

Cutoff Frequency

$$\omega_c = \frac{2}{\sqrt{LC}} \tag{3-35}$$

Attenuation for $0 \le \omega \le \omega_c$)

$$\alpha = 0$$

The phase is

$$\theta = 2(n) \sin^{-1} \left(\frac{\omega}{\omega_c} \right) \tag{3-36}$$

Attenuation for $\omega_c \leq \omega$

$$\alpha = 2(n) \cosh^{-1} \left| \frac{\omega}{\omega_c} \right| \qquad (3\text{-}37)$$

The phase is

$$\theta = \pi n$$

n is the number of sections.

The mutual inductance is assumed to be zero.

Capacitor-Input Alternate Form

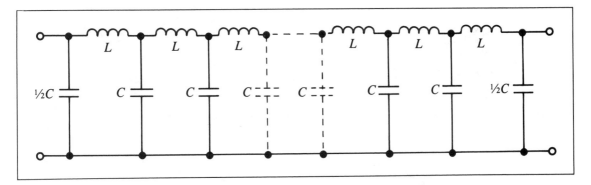

Impedance when $\omega \ll \omega_e$

$$Z_o = \sqrt{\frac{L}{C}} \qquad (3\text{-}38)$$

Cutoff Frequency

$$f_c = \frac{1}{\pi\sqrt{LC}} \qquad (3\text{-}39)$$

L is inductance of one coil (in *henries*)
C is capacitance (in *farads*)

Time Delay for Low Frequencies

$$t_d = n\sqrt{LC} \qquad (3\text{-}40)$$

n is the number of sections

WIRE LINE

A wire line replaces the strip line with insulated wire in close proximity to a ground plane.

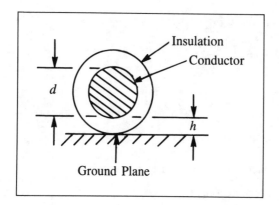

Impedance

$$Z_o = \frac{138}{\sqrt{e}} \log_{10} \frac{4\,h}{d} \qquad\qquad (3\text{-}41)$$

e is the dielectric constant of the insulation

MICROSTRIP LINE

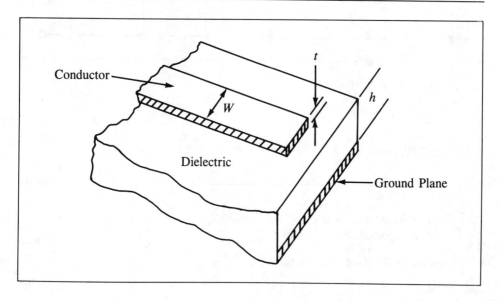

PHASE VELOCITY

$$V_p = \frac{c}{\sqrt{\epsilon_{\text{eff}}}} \qquad\qquad (3\text{-}42)$$

WAVELENGTH IN THE LINE

$$\lambda = \frac{V_p}{f}$$ (3-43)

IMPEDANCE

$$Z_o = \frac{1}{V_p C}$$ (3-44)

where

    c is the velocity of light
    ϵ_{eff} is the effective dielectric constant of the substrate
    f is frequency
    C is capacitance per unit length of line

MICROSTRIP IMPEDANCE

$$Z_o = \frac{377\,h}{\epsilon_r^{\frac{1}{2}} W_{eff}\left[1 + \left(1.735\,\epsilon_r\right) - \left[0.0724\left(\frac{W_{eff}}{h}\right)\right] - 0.836\right]}$$ (3-45)

where

    ϵ_r is the dielectric constant of the substrate
    h is the thickness of the dielectric board material
    $W_{eff} = W + (t/\pi)[\ln(2h/t) + 1]$
        W is the width of the microstrip line
        t is the thickness of the strip conductor

MICROSTRIP IMPEDANCE—SOBOL'S EQUATION

$$Z_o = \frac{120\pi h}{\sqrt{e_r}\,w(1 + 1.735\,e_r^{-0.0724}\,w/h^{-0.836})}$$ (3-46)

where

    Z_o is the impedance
    w is the width of the microstrip
    h is the thickness of the dielectric material
    e_r is the relative dielectric constant of the substrate

When $e_r > 4.0$,

$$Z_o \approx \frac{120\pi}{(w/h + 1)\sqrt{(e_r + \sqrt{e_r})}}$$ (3-47)

$$\frac{w}{h} \approx \frac{120\pi}{Z_o\sqrt{(e_r + \sqrt{e_r})}} - 1$$ (3-48)

HAMMERSTAD'S EQUATIONS $(t \to 0)$

If $W/h \leq 1$,

$$Z_o = \frac{60}{\sqrt{\epsilon_{\text{eff}}}} \ln(8h/W + 0.25W/h) \tag{3-49}$$

where

$$\epsilon_{\text{eff}} = \frac{\epsilon_r + 1}{2} + \frac{\epsilon_r - 1}{2} [(1 + 12h/W)^{-\frac{1}{2}} + 0.04(1 - W/h)^2] \tag{3-50}$$

If $W/h \geq 1$

$$Z_o = \frac{120\pi/\sqrt{\epsilon_{\text{eff}}}}{W/h + 1.393 + 0.667 \ln(W/h + 1.444)} \tag{3-51}$$

where

$$\epsilon_{\text{eff}} = \frac{\epsilon_r + 1}{2} + \frac{\epsilon_r - 1}{2}(1 + 12 \, h/W)^{-\frac{1}{2}} \tag{3-52}$$

Accuracy for $0.05 \leq W/h \leq 20$ and $\epsilon_r \leq 16$

$\epsilon_{\text{eff}} = \pm\frac{1}{2}\%; \quad Z_0 = 0.8\%$

Useful Range

$2 \leq \epsilon_r \leq 10; \quad t/h \leq 0.005; \; 0.1 \leq W/h \leq 5$

When $W/h \leq 2$

$$W/h = \frac{8e^A}{e^{2A} - 2} \tag{3-53}$$

When $W/h \geq 2$

$$W/h = \frac{2}{\pi}[B - 1 - \ln(2B - 1) + C] \tag{3-54}$$

$$A = \frac{Z_o}{60} \sqrt{\frac{\epsilon_r + 1}{2}} + \frac{\epsilon_r - 1}{\epsilon_r + 1}(0.23 + 0.11/\epsilon_r) \tag{3-55}$$

$$B = \frac{377\pi}{2 \, Z_0\sqrt{\epsilon_r}} \tag{3-56}$$

$$C = \frac{\epsilon_r - 1}{2\epsilon_r}\left[\ln(B - 1) + 0.39 - \frac{0.61}{\epsilon_r}\right] \tag{3-57}$$

When t is 0.0002 to 0.0005-inch metalized alumina or t is 0.001 inch to 0.003 inch low dielectric material:

If $W/h \geq \frac{1}{2}\pi$

$$W_{\text{eff}}/h = W/h + \frac{t}{\pi h}\left(1 + \ln\frac{2h}{t}\right) \tag{3-58}$$

If $W/h \leq \frac{1}{2}\pi$

$$W_{\text{eff}}/h = W/h + \frac{t}{\pi h}\left(1 + \ln\frac{4\pi W}{t}\right) \tag{3-59}$$

MICROSTRIP EFFECTIVE WAVELENGTH

$$\lambda = \frac{\lambda_o}{\sqrt{\epsilon_r}}\left[\frac{\epsilon_r}{1 + 0.63(\epsilon_r - 1)\left(\dfrac{W_{\text{eff}}}{h}\right)^{0.1225}}\right]^{\frac{1}{2}} \tag{3-60}$$

(See Microstrip Impedance for term definitions)

$\lambda_o = c/f_o$
c is the speed of light in a vacuum or free space
f_o is the frequency at mid band

Note: Equation as shown is usable for $W_{\text{eff}}/h > 0.6:1$. *For W_{eff}/h ratios that are* $<0.6:1$, change the exponent from 0.1225 to 0.0279.

MICROSTRIP CORNER

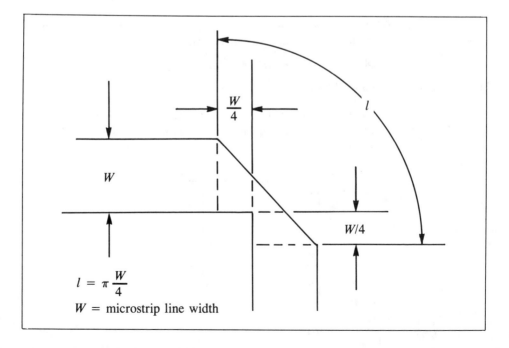

WAVEGUIDES

WAVEGUIDE CUTOFF FREQUENCY (f_c)

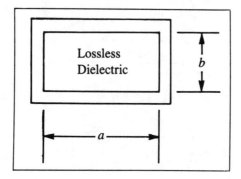

For TE$_{mn}$ and TM$_{mn}$ moles,

$$f_c = \frac{c}{2}\left[\left(\frac{m}{a}\right)^2 + \left(\frac{n}{b}\right)^2\right]^{\frac{1}{2}} \quad \text{(in } GHz\text{)} \tag{3-61}$$

$$c = 299.8/\sqrt{\epsilon_r} \quad \text{(in } mm \text{ per } ns\text{)} \tag{3-62}$$

ϵ_r is the relative permittivity
m and n are integers that define the mode. See Table 3-2.

Dimensions are in mm.

Note: The permittivity is equal to the product of the relative dielectric constant times the permittivity of a vacuum. The permittivity of a vacuum is

$$10^{-9}/36\pi = 8.884 \cdot 10^{-12} \text{ farads per meter}$$

WAVEGUIDE DOMINANT MODE

The dominant mode of operation is the most used because of efficiency considerations. If a is in inches, the approximate frequencies of operation for this mode in any air dielectric waveguide can be calculated from:

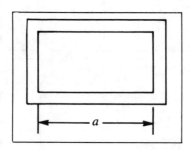

$$F_{\min} = \frac{7376}{a} \quad \text{(in } MHz\text{)} \tag{3-63}$$

$$F_{max} = \frac{11,136}{a} \quad \text{(in } MHz)} \tag{3-64}$$

WAVEGUIDE WAVELENGTH

$$\lambda_g = \frac{\lambda}{K^{\frac{1}{2}}[1 - (\lambda/\lambda_c)^2]^{\frac{1}{2}}} \tag{3-65}$$

λ is the free-space wavelength
K is the dielectric constant (air = 1)
$\lambda_c = c/f_c$
f_c is the waveguide cutoff frequency
c is the velocity of light

WAVEGUIDE ATTENUATION BELOW CUTOFF

$$\alpha = \frac{2\pi}{c}[f_c^2 - f^2]^{\frac{1}{2}} \quad \text{(in } N_p/mm) \tag{3-66}$$

$$c = 299.8/\sqrt{\epsilon_r} \quad \text{(in } mm/ns) \tag{3-67}$$

f_c is the cutoff frequency (in GHz)
f is the frequency of interest (in GHz)

$$\alpha = \frac{2\pi \, 8686}{c}[f_c^2 - f^2]^{\frac{1}{2}} \quad \text{(in } dB/m) \tag{3-68}$$

ϵ_r is the relative permittivity

WAVEGUIDE WAVELENGTH (λ_g) ABOVE CUTOFF (f_c)

$$\lambda_g = \frac{c}{[f^2 - f_c^2]^{\frac{1}{2}}} \quad \text{(in } mm) \tag{3-69}$$

$c = 299.8/\sqrt{\epsilon r}$ (in mm/ns)
ϵ_r is the relative permittivity

Frequencies are in GHz.

CIRCULAR WAVEGUIDES

Table 3-2. Cutoff Wavelength (λ_c)

Mode	λ_c
TE$_{0,1}$	0.82d
TM$_{0,1}$	1.31d
TE$_{1,1}$	1.71d
TM$_{1,1}$	0.82d
TE$_{2,1}$	1.03d
	d is diameter

MATCHING

DOUBLE QUARTER-WAVE SECTIONS FOR MATCHING

This type of matching greatly improves the bandwidth of the match over that of a single section.

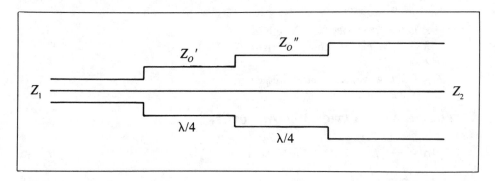

$$(Z_o')^4 = (Z_1)^3(Z_2) \qquad (3\text{-}70)$$
$$(Z_o'')^4 = (Z_1)(Z_2)^3 \qquad (3\text{-}71)$$

Z_1 is to be matched to Z_2
Z_o' is the section next to Z_1
Z_o'' is the section next to Z_2

IMPEDANCE TRANSFORMATION USING TRANSMISSION LINES

Given:
- L transmission line length (in *cm*)
- f frequency (in *Hz*)
- v propagation velocity of the line
- Z characteristic line impedance (in *ohms*)
- Z_L termination impedance (in *ohms*)

electrical length (in *degrees*)

$$\theta = 1.20083 \cdot 10^{-8} \, fL\sqrt{\epsilon_r} \qquad (3\text{-}72)$$

Input Impedance

$$Z_{in} = Z\left[\frac{(Z_L/Z) + j \tan \theta}{1 + j(Z_L/Z) \tan \theta}\right] \qquad (3\text{-}73)$$

ϵ_r is the dielectric constant of the line insulator

TRANSMISSION LINE IMPEDANCE MATCHING

$$Z_o = \left(\frac{1}{R_{in} - R_{out}}\left[(R_{in}^2 + X_{in}^2)R_{out} - (R_{out}^2 + X_{out}^2)R_{in}\right]\right)^{\frac{1}{2}} \qquad (3\text{-}74)$$

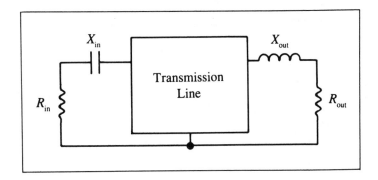

$$\theta = \tan^{-1}\left[Z_o\frac{R_{\text{out}} - R_{\text{in}}}{X_{\text{in}} R_{\text{out}} - X_{\text{out}} R_{\text{in}}}\right] \tag{3-75}$$

where

Z_o is the impedance of the line
θ is the phase delay of the line
l is the length of the line as in:

$$l = \theta\,\frac{V_p}{600\,f} \quad (\text{in } cm) \tag{3-76}$$

V_p is the velocity of propagation of the line (in *percent*)
f is frequency (in *GHz*)

4

---

filters

COMPOSITE FILTERS

A filter can be composed by combining the constant k and m-derived filter sections to secure the characteristic best suited to the application using appropriate terminating half-sections.

CONSTANT-k FILTER SECTIONS

$$k = (Z_1 \cdot Z_2)^{\frac{1}{2}} = R \tag{4-1}$$

where

k is a constant at all frequencies
R is the terminating resistance
Z_1 and Z_2 are the series and shunt impedances of the filter elements

m-DERIVED FILTER SECTIONS

Characteristics compared to constant k: Sharper cutoff (f_c)
Lower attenuation $(f > f_c)$

Types

Series derived—the shunt arm is resonant
Shunt derived—the series arm is resonant

m Defined

$$m = \left[1 - \left(\frac{f_c}{f_\infty} \right)^2 \right]^{\frac{1}{2}} \tag{4-2}$$

where

f_c is the cutoff frequency
f_∞ is the frequency of infinite attenuation
$0 < m < 1$

Effect of *m*

As $m \rightarrow 0$, the cutoff sharpness increases, sacrificing the attenuation at $f > f_c$.

Compromise Value of *m*

A value of $m = 0.6$ is a good compromise value for design use.

Frequency of Infinite Attenuation (f_∞)

The frequency of infinite attenuation is determined by the resonant arm of the filter.

The amount of attenuation at the frequency of infinite attenuation is determined by the loaded Q of the resonant elements.

CONFIGURATIONS

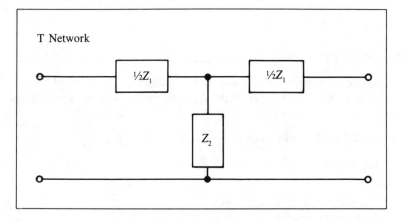

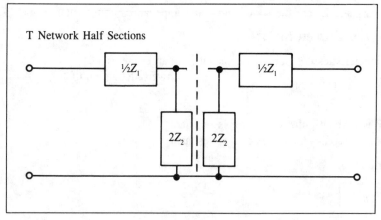

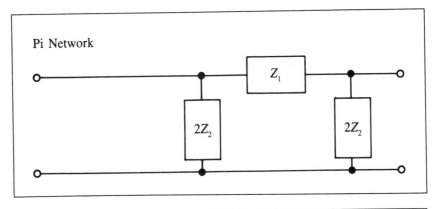

Pi Network

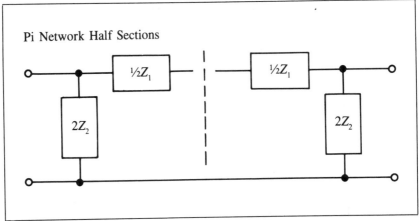

Pi Network Half Sections

m-DERIVED FILTERS

Low-Pass, *m*-Derived Filter Sections

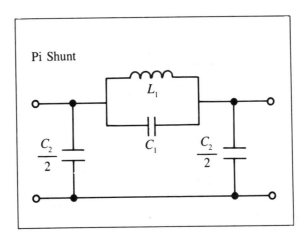

Pi Shunt

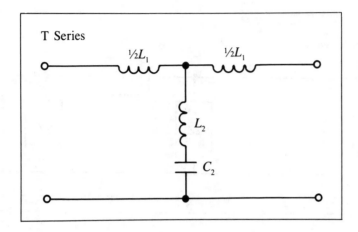

$$L_1 = mL \tag{4-3}$$
$$L = R/(\pi f_c) \tag{4-4}$$
$$L_2 = (1 - m^2)L/(4m) \tag{4-5}$$
$$C_2 = mC \tag{4-6}$$
$$C_1 = C(1 - m^2)/(4m) \tag{4-7}$$
$$C = 1/(\pi R f_c) \tag{4-8}$$

R is the source and load resistance

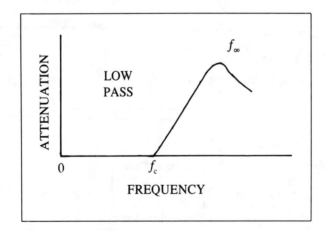

Low-Pass, *m*-Derived, Terminating Half Sections

For value definitions, see equations (4-3) through (4-8).

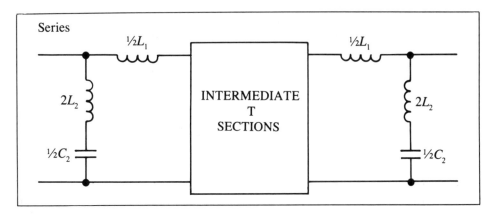

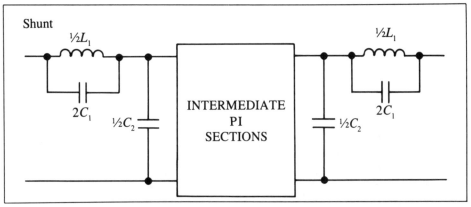

High-Pass, *m*-Derived Filter Sections

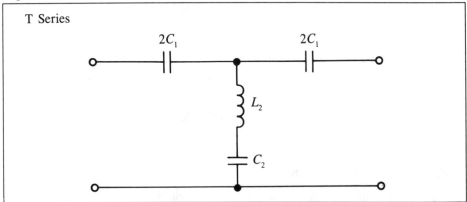

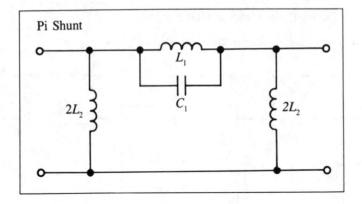

$$L = R/(4\pi f_c) \tag{4-9}$$
$$L_1 = 4\, mL/(1 - m^2) \tag{4-10}$$
$$L_2 = L/m \tag{4-11}$$
$$C = 1/(4\pi f_c R) \tag{4-12}$$
$$C_1 = C/m \tag{4-13}$$
$$C_2 = 4mC/(1 - m^2) \tag{4-14}$$

R is the source and load resistance

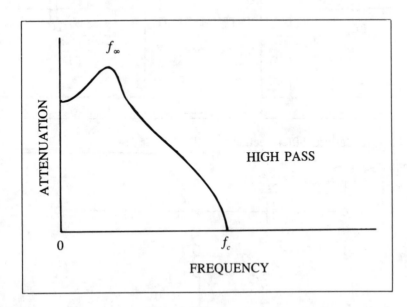

High Pass, *m*-Derived, Terminating Half Sections

For value definitions, see equations (4-9) through (4-14).

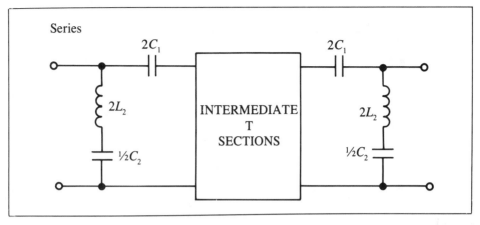

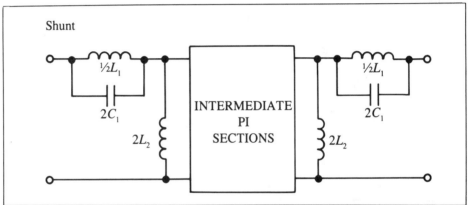

Shunt, *m*-Derived, Bandpass Filter Sections

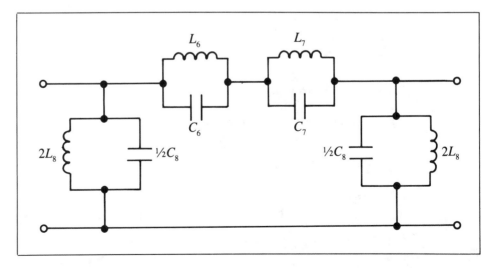

$$L_6 = mL_1 \left[\frac{(1 + 1/N)^2}{1 + N^2} \right] \tag{4-15}$$

$$L_8 = L_2/m \tag{4-16}$$

$$C_6 = \frac{C_1}{m} \left[\frac{1 + 1/N^2}{(N - 1/N)^2} \right] \tag{4-17}$$

$$C_8 = mC_2 \tag{4-18}$$

$$L_7 = mL_1 \left[\frac{(N - 1/N)^2}{1 + 1/N^2} \right] \tag{4-19}$$

$$C_7 = \frac{C_1}{m} \left[\frac{1 + N^2}{(N - 1/N)^2} \right] \tag{4-20}$$

where

$$C_1 = \frac{f_2 - f_1}{4\pi f_1 f_2 R}$$

$$L_1 = \frac{R}{\pi(f_2 f_1)}$$

$$C_2 = \frac{1}{\pi(f_2 - f_1)R}$$

$$L_2 = \frac{R(f_2 - f_1)}{4\pi f_1 f_2}$$

(The above equations are repeated in equations (4-34) through (4-37). The variables are defined subsequently.)

$$N = f_{2\infty}/f_m \tag{4-21}$$

$$m = \left[1 - \frac{(f_2/f_m - f_m/f_2)^2}{(f_{2\infty}/f_m - f_m/f_{2\infty})^2} \right]^{\frac{1}{2}} \tag{4-22}$$

$$f_m = (f_1 f_2)^{\frac{1}{2}} = (f_{1\infty} f_{2\infty})^{\frac{1}{2}} \tag{4-23}$$

where

f_1 is the lower cutoff frequency
f_2 is the upper cutoff frequency
$f_{1\infty}$ is the lower frequency of infinite attenuation
$f_{2\infty}$ is the upper frequency of infinite attenuation

Terminating Half Section: Bandpass Case

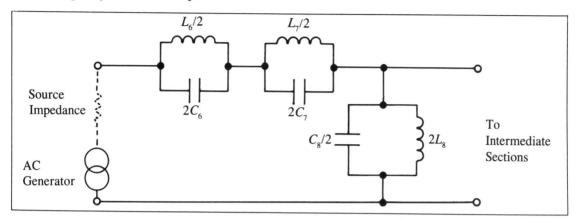

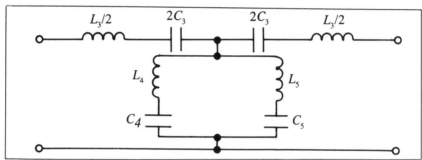

Series, *m*-Derived, Bandpass Filter Sections

$$L_3 = mL_1 \tag{4-24}$$

$$L_4 = L_1 A(1 + 1/N^2) \tag{4-25}$$

$$C_3 = C_1/m \tag{4-26}$$

$$C_4 = \frac{1}{A}\left[\frac{C_1}{1 + N^2}\right] \tag{4-27}$$

$$L_5 = L_1 A(1 + N^2) \tag{4-28}$$

$$C_5 = \frac{1}{A}\frac{C_1}{1 + 1/N^2} \tag{4-29}$$

where

$A = (1 - m^2)/4m$

N is defined in equation (4-21)

m is defined in equation (4-22)

Equations (4-34) through (4-37) in the next section list the formulas for calculating C_1, C_2, L_1, and L_2, respectively.

Terminating Half Section—Bandpass Case

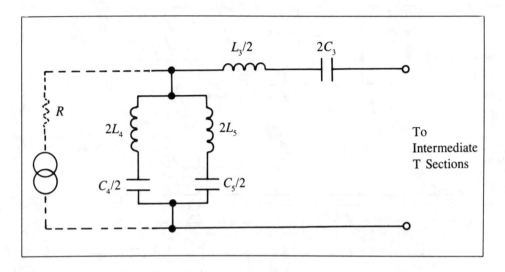

CONSTANT-*k* FILTER SECTIONS

Low-Pass Constant-*k* Sections

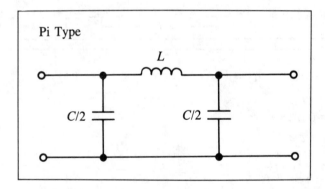

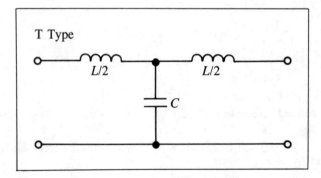

$$L = \frac{R}{\pi f_c} \tag{4-30}$$

$$C = \frac{1}{\pi R f_c} \tag{4-31}$$

R is the source and load resistance

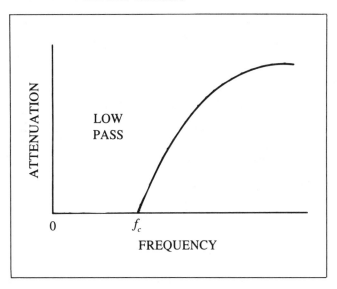

High-Pass Constant-k Sections

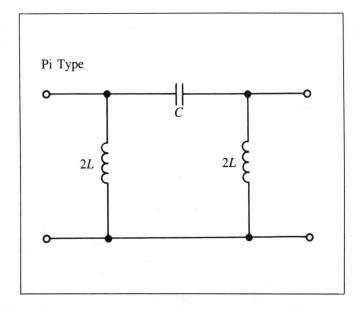

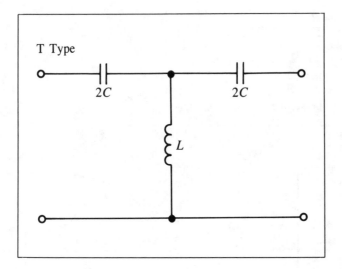

$$L = \frac{R}{4\pi f_c} \qquad (4\text{-}32)$$

$$C = \frac{1}{4\pi f_c R} \qquad (4\text{-}33)$$

R is the source and load resistance

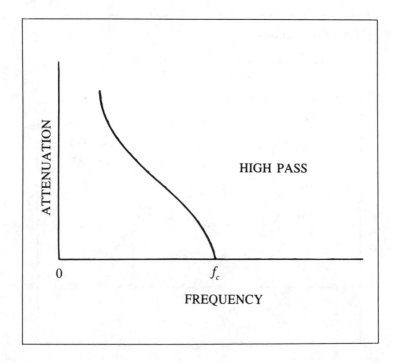

Bandpass Constant-k Sections

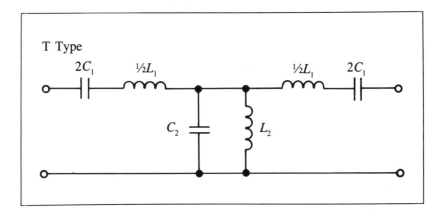

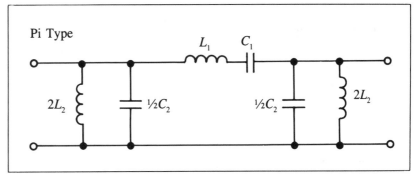

$$C_1 = \frac{f_2 - f_1}{4\pi f_1 f_2 R} \tag{4-34}$$

$$L_1 = \frac{R}{\pi(f_2 - f_1)} \tag{4-35}$$

$$C_2 = \frac{1}{\pi(f_2 - f_1)R} \tag{4-36}$$

$$L_2 = \frac{R(f_2 - f_1)}{4\pi f_1 f_2} \tag{4-37}$$

where
f_1 is the lower -3 dB point
f_2 is the upper -3 dB point
R is the load
C is capacitance in farads
L is inductance in henries

BUTTERWORTH FILTERS

BUTTERWORTH LOW-PASS FILTER COMPONENTS

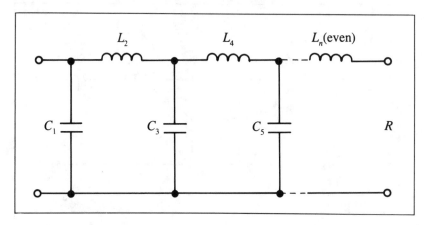

Where i is odd and equal to or less than n,

$$C_i = \frac{1}{\pi f_c R} \sin \frac{(2i - 1)\pi}{2n}$$

(4-38)

Where i is even and equal to or less than n,

$$L_i = \frac{R}{\pi f_c} \sin \frac{(2i - 1)\pi}{2n}$$

(4-39)

f_c is the cutoff frequency
R is the source and load resistance
L_n is the final inductor in the series

BUTTERWORTH FILTER FREQUENCY RESPONSE

$$A = 10 \log_{10}(1 + k^{2n})$$

(4-40)

where
   A is the response of the filter (in dB)
   $k = b_x/b_3$ for bandpass and low-pass filters
   $k = b_3/b_x$ for high-pass and notch filters
   b_x is the bandwidth at frequency x
   b_3 is the bandwidth at the -3 dB points
   n is the number of reactances for low- and high-pass filters
   n is the number of resonators for band-pass and notch (band-reject) filters

BUTTERWORTH FILTER GROUP DELAY

$$D_g = \frac{1}{2\pi(f_{max} - F_{min})} (1 + J) \sum_{k=1}^{n} \frac{|\sigma_k|}{(\sigma_k)^2 + (x - \omega_k)^2}$$

(4-41)

where

D_g is group delay (in *seconds*)

f_{max} is the upper -3 dB frequency

f_{min} is the lower -3 dB frequency

$J = 0$ for the low-pass case and 1 for the band-pass case

n is the number of reactances for the low-pass case and the number of resonators for the band-pass case

$\sigma_k = \cos \theta_k$

$\omega_k = \sin \theta_k$

$$\theta_k = \frac{90°(2k + n - 1)}{n} \tag{4-42}$$

$$x = \left| \frac{1}{f_{max} - f_{min}} \left(f - \frac{f_{max}f_{min}}{f} \right) \right| \tag{4-43}$$

f is the variable frequency

All frequency entries are in Hz.

CHEBYSHEV FILTERS

CHEBYSHEV LOW-PASS FILTER COMPONENTS

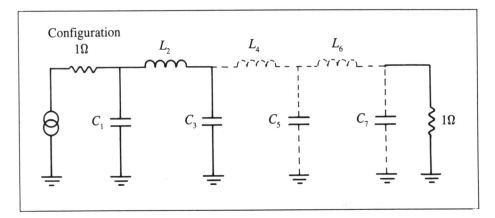

Transposition to Actual Values

$$C_m' = \frac{C_m}{R\omega_c} \tag{4-44}$$

$$L_m' = \frac{RL_m}{\omega_c} \tag{4-45}$$

$$\omega_c = 2\pi f_c \tag{4-46}$$

f_c is the desired cutoff frequency (in *Hz*)

R is the desired source and load resistances (in *ohms*)

Table 4-1. Normalized Values of L and C for $n = 3$, 5, and 7

Ripple values are shown for 0.1, 0.25, 0.5, and 1 dB cases.
R_{source} and R_{load} are equal to 1 ohm.
L is in henries and C is in farads.

$n = 3$	C_1, C_3	L_2		
0.1	1.032	1.147		
0.25	1.303	1.146		
0.5	1.596	1.097		
1.0	2.024	0.994		
$n = 5$	C_1, C_5	L_2, L_4	C_3	
0.1	1.147	1.371	1.975	
0.25	1.382	1.326	2.209	
0.5	1.706	1.230	2.541	
1.0	2.135	1.091	3.001	
$n = 7$	C_1, C_7	L_2, L_6	C_3, C_5	L_4
0.1	1.181	1.423	2.097	1.573
0.25	1.447	1.356	2.348	1.469
0.5	1.737	1.258	2.638	1.344
1.0	2.167	1.112	3.094	1.174

Alternate Form (using the table values)

Change the series inductors to shunt capacitors. Then

$$C = L/R^2 \qquad (4\text{-}47)$$

Change the shunt capacitors to series inductors. And

$$L = CR^2 \qquad (4\text{-}48)$$

EXAMPLE:

Given, using the standard form:
  Cutoff frequency = 10 MHz
  Source and load resistances = 50 ohms
  Ripple = 0.5 dB
  The number of components (n) = 3

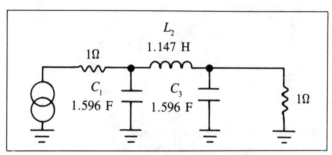

$$C_1' = C_3' = \frac{1.596}{50 \cdot 2\pi 10^7} = 508 \text{ pF} \tag{4-49}$$

$$L_2' = \frac{1.147 \cdot 50}{2\pi 10^7} = 0.9135 \text{ }\mu\text{H} \tag{4-50}$$

CHEBYSHEV FILTER FREQUENCY RESPONSE

$$A = 10 \log_{10} \left\{ 1 + \left[\left(\log^{-1} \frac{A_{max}}{10} \right) - 1 \right] \cosh^2 \left[n \cdot \cosh^{-1}(k) \right] \right\} \tag{4-51}$$

A is the response of the filter (in dB)

A_{max} is the ripple of the filter (in dB)

n is the number of reactances for the low- and high-pass cases and is the number of resonators for the band-pass and notch filters

$k = b_3/b_x$ for the high-pass and band-reject cases

$k = b_x/b_3$ for the band-pass and low-pass cases

b_x is the bandwidth at a frequency x

b_3 is the -3 dB bandwidth

$k > 2$

CHEBYSHEV FILTER GROUP DELAY

$$D_g = \frac{1}{2\pi(f_{max} - f_{min})} \left[1 + \frac{f_{max}f_{min}}{f^2} \right] \sum_{k=1}^{n} \frac{|\sigma_k|}{(\sigma_k)^2 + (\Omega - \omega_k)^2} \tag{4-52}$$

where

D_g is group delay (in *seconds*)

f_{max} is the upper -3 dB frequency

f_{min} is the lower -3 dB frequency

f is the variable frequency at which the delay is desired

n is the number of reactances for the low- and high-pass case or the number of resonators for the band-pass and notch (band-reject) case

$$\sigma_k = \sinh \left[\frac{1}{n} \sinh^{-1} \frac{1}{a} \right] \sin \theta_k \tag{4-53}$$

$$\omega_k = \cosh \left[\frac{1}{n} \sinh^{-1} \frac{1}{a} \right] \cos \theta_k \tag{4-54}$$

$$a = \left[\log^{-1} \left(\frac{A_{max}}{10} \right) - 1 \right]^{\frac{1}{2}} \tag{4-55}$$

A_{max} is ripple (in dB)

$$\Omega = \left| \frac{1}{f_{max} - f_{min}} \left(f - \frac{f_{max}f_{min}}{F} \right) \right| \tag{4-56}$$

$$\sinh^{-1} x = \ln \left[x + (x^2 + 1)^{\frac{1}{2}} \right] \tag{4-57}$$

$$\sinh x = \tfrac{1}{2}(e^x - e^{-x}) \tag{4-58}$$

$$\cosh x = \tfrac{1}{2}(e^x + e^{-x}) \tag{4-59}$$

All frequency entries are in Hz.

ACTIVE FILTERS

ACTIVE LOW-PASS FILTER

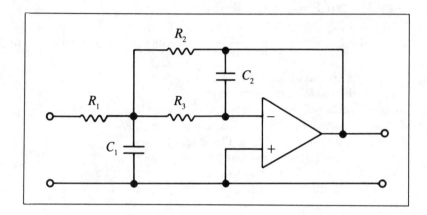

$$R_1 = R_2/G \qquad (4\text{-}60)$$

$$R_2 = \alpha/(4\pi f_c C_2) \qquad (4\text{-}61)$$

$$R_3 = R_2/(G + 1) \qquad (4\text{-}62)$$

$$C_1 = \frac{4(1 + G)C_2}{\alpha^2} \qquad (4\text{-}63)$$

where

G is the passband gain of values 10 to 100 for peaking factors $\alpha = 0.1$ to 2.0
α is the peaking factor (0.1 to 2.0)
f_c is the cutoff frequency (in Hz)
C is capacity in farads

The open loop gain should be 75 to 100 dB
Note: The peaking factor determines the amount of overshoot at the cutoff frequency. For maximum flatness, a value of $\sqrt{2}$ should be used. The lower the peaking factor, the more the overshoot.

HIGH-PASS ACTIVE FILTER

$$C_1 = C_3 \qquad (4\text{-}64)$$

$$C_2 = C_1/G \qquad (4\text{-}65)$$

$$R_1 = \frac{2G + 1}{2\pi f_c C_1 \alpha} \qquad (4\text{-}66)$$

$$R_2 = \frac{\alpha}{2\pi f_c C(2 + 1/G)} \qquad (4\text{-}67)$$

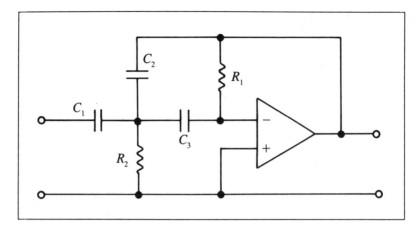

where
    C is in farads
    G is passband gain
    α is the peaking factor (equals 0.1 for $G < 10$; equals 1.0 for $G = 100$)
    f_c is the cutoff frequency (in Hz)

The open-loop gain of the amplifier should be 75 to 100 dB.

ACTIVE BANDPASS FILTERS

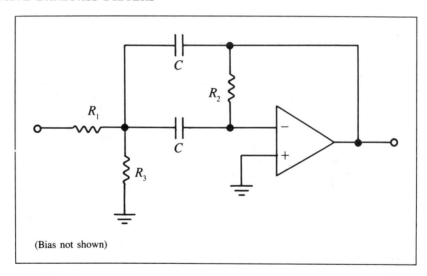

(Bias not shown)

$$R_1 = \frac{1}{2\pi ABC} \quad \text{(select } C\text{)} \qquad (4\text{-}68)$$

$$R_2 = 2R_1A = \frac{1}{\pi BC} \qquad (4\text{-}69)$$

$$R_3 = \frac{1}{2\pi C}\left[\frac{1}{2f^2/B) - (B/A)}\right]$$ (4-70)

$$R_3 = \left\{\frac{1}{2\pi C[(2f^2/B) - BA]}\right\}^{\frac{1}{2}}$$ (4-71)

A is gain
B is bandwidth (in Hz)
f is center frequency

Alternate Form

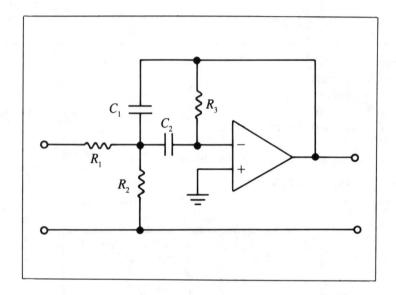

$$R_1 = Q/(G\omega_o C_1)$$ (4-72)

$$R_2 = \frac{R_1 k}{R_1 - k}$$ (4-73)

$$R_3 = \frac{Q(C_1 + C_2)}{C_1 C_2 \omega_o}$$ (4-74)

$$k = \frac{1}{Q\omega_o(C_1 + C_2)}$$ (4-75)

$Q > .707G^{\frac{1}{2}}$ (choose Q and G)

where
$\omega_o = 2\pi f$
f is the center frequency
G is gain
C is in farads (choose C_1 and C_2)

ACTIVE NOTCH FILTER

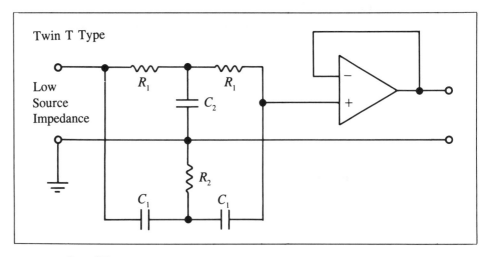

Twin T Type

Low Source Impedance

$R_1 = 2R_2$

$C_1 = \frac{1}{2}C_2$

$$f_o = \frac{1}{2\pi R_1 C_1}$$

(4-76)

$Q \approx 0.3$

The attenuation approaches infinity when driven from a low impedance and loaded by a high impedance if the components are well matched.

The Q of the circuit can be increased to over 50 by providing feedback to the $R_2 C_2$ junction from the output as shown in the following figure.

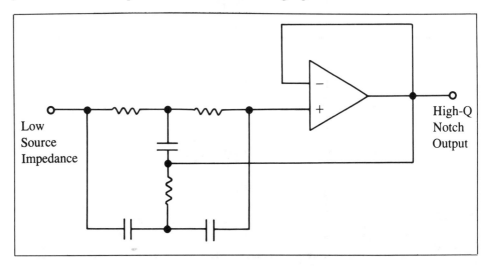

Low Source Impedance

High-Q Notch Output

The depth of the notch can be controlled by adding a potentiometer between the input and the output and taking the output from the wiper arm.

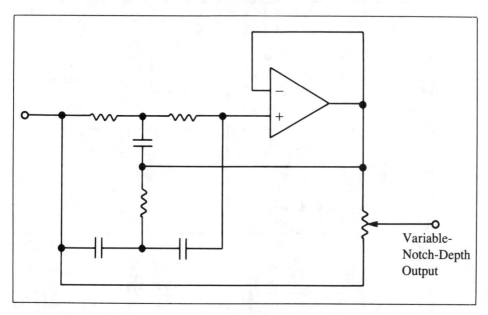

The Q of the circuit can be varied by reducing the feedback to R_2C_2. This allows the control of Q over a range of about 0.3 to more than 50.0. The R_2C_2 junction must be fed from a low-impedance source as shown in the following circuit.

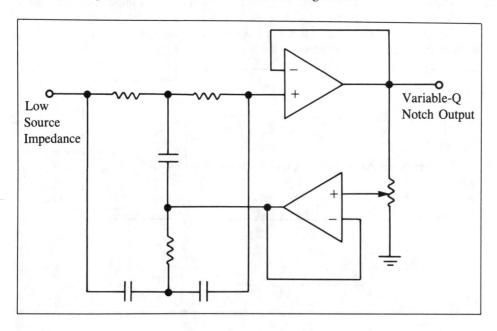

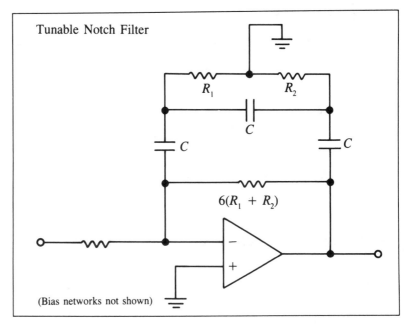

$$f_{\text{null}} = \frac{1}{2\pi C(3R_1 R_2)^{\frac{1}{2}}}$$

(4-77)

NONMATCHED FILTERS

B is bandwidth and T is pulse width.

Table 4-2. Gaussian Pulse Input

Filter	BT Optimum	Loss Constant
Gaussian	0.44	1.0
Rectangular	0.72	0.893

Table 4-3. Rectangular Pulse Input

Filter	BT Optimum	Loss Constant
RC	0.199	0.816
RLC	0.398	0.816
n (RLC) where n ≥ 2	0.61 to 0.68	0.879 to 0.891
Gaussian	0.72	0.893
Rectangular	1.37	0.822
n (RLC) is the number of RLC stages		

NONMATCHED RECEIVER DETECTOR FILTERING

RC Low-Pass Filter Efficiency

$$\eta_{RC} = \frac{2RC}{\tau} \left[1 - \exp\left(-\frac{\tau}{RC}\right) \right]^2 \tag{4-78}$$

When $\tau/RC = 1.25$, η_{RC} is maximized. Further,

$$B_{\frac{1}{2}}\tau = \frac{1}{2\pi RC} \tag{4-79}$$

τ is the width of the rectangular pulse

$B_{\frac{1}{2}}$ is the $\frac{1}{2}$-power bandwidth of the filter

$$B_{\frac{1}{2}}\tau = 0.2 \tag{4-80}$$

The S/N degradation is 0.883 dB compared to a matched filter.

5

baluns, directional couplers, and dividers

BRIDGE BALUN

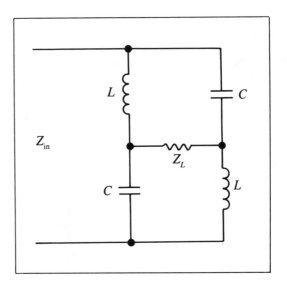

$$Z_{in} = \frac{2\omega L + j(k^2 - 1)Z_L}{2\omega C\, Z_L + j(k^2 - 1)} \tag{5-1}$$

where

$$k = \omega/\omega_o \tag{5-2}$$
$$\omega_o = 1/LC \quad \text{(band center)} \tag{5-3}$$

When $k = 1$

$$L = (Z_{in}Z_L)^{\frac{1}{2}}/\omega \tag{5-4}$$
$$C = 1/[\omega(Z_{in}Z_L)^{\frac{1}{2}}] \tag{5-5}$$

HALF-WAVE COAXIAL BALUN

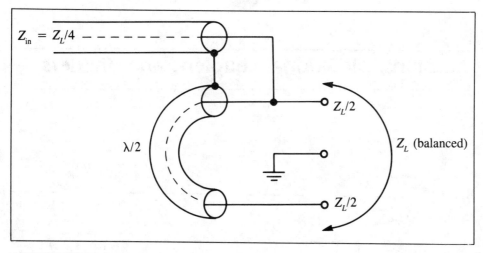

DIRECTIONAL COUPLERS

QUADRATURE COUPLERS (LUMPED CONSTANT)

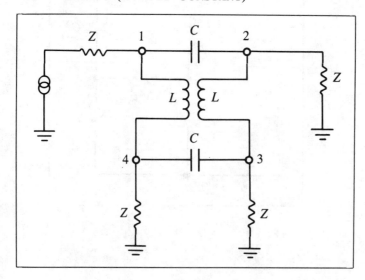

Input Port   1
Output Port   2, 4
Phase $|\theta_2 - \theta_1| = 90°$ (5-6)

The minimum loss case results when $f_o = f$ and

$$\frac{P_1}{P_2} = \frac{P_1}{P_4} = 2 = 3 \text{ dB loss} \tag{5-7}$$

where

P is the power level at the port.

Relationships for Coupling with Any Loss

$$\frac{P_1}{P_2} = 1 + (f_o/f)^2 \tag{5-8}$$

or

$$\frac{P_1}{P_4} = 1 + (f/f_o)^2 \tag{5-9}$$

where

f is the frequency of operation
f_o is the -3 dB coupling frequency

$$Z = (L/C)^{\frac{1}{2}} \tag{5-10}$$

$$= \frac{1}{2\pi f_o C} \tag{5-11}$$

$$= 2\pi f_o L \tag{5-12}$$

EXAMPLE:

Let $P_1/P_2 = 10$ at 14 MHz and a Z of 50 ohms. Then, from equation (5-8),

$$(f_o/f)^2 = 9$$

and

$f_o = 3f = 42$ MHz
$L = Z_o/\omega_o$
$C = 1/(\omega_o Z_o)$
$L = 50/2\pi 14 \cdot 10^6 = 0.568 \ \mu H$
$C = 1/2\pi 14 \cdot 10^6 \cdot 50 = 227.36$ pF

Also,

$P_1/P_4 = 1 + (14/42)^2 = 1.111$
$P_4 = 0.9 \ P_1 = 0.457$ dB

Construction

L is twisted varnished wire such as #30 with 10 twists per inch wound on an air core (VHF) or a torroidal ferrite core for HF.

Low-Frequency Directional Coupler

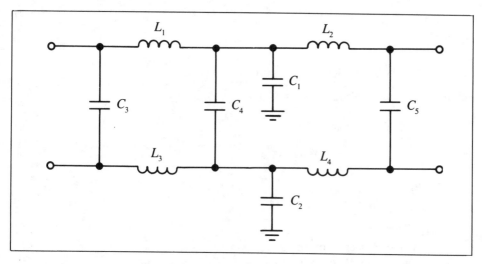

$$L_n = Z_o/\omega \quad \text{for } n = 1 \text{ through } 4 \tag{5-13}$$

$$Z_o = 1/(\omega C_1) = 1/(\omega C_2) \tag{5-14}$$

$$C_3 = C_4 = C_5 = C_1/2 \tag{5-15}$$

High-Frequency Coupler

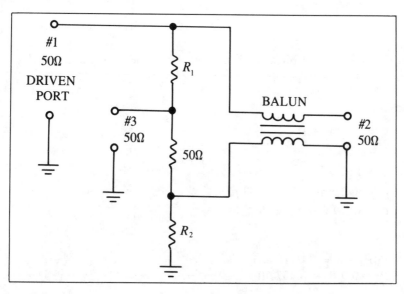

Isolation Ports: 2 and 3

$$I = (R_1 + 50)/(R_2 + 50) \tag{5-16}$$

Loss Port: 2 Relative to 1

$$L = 50/(50 + R_2) \tag{5-17}$$

where

$$R_1 = 2500/R_2 = 50I\Omega \tag{5-18}$$

$$R_2 = \frac{-50\left(\dfrac{I-1}{I}\right) \pm \left\{\left[-50\left(\dfrac{I-1}{I}\right)\right]^2 + \dfrac{10000}{I}\right\}^{\frac{1}{2}}}{2} = \frac{50}{I}\Omega \tag{5-19}$$

EXAMPLE:

$$I = 10, \ R_1 = 500\Omega, \ R_2 = 5\Omega$$

DIVIDERS

n-WAY DIVIDERS

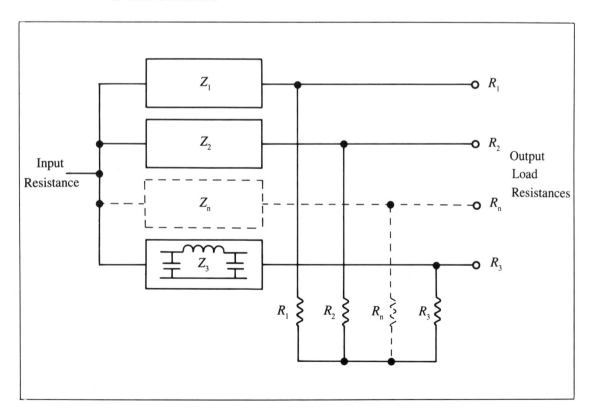

where
$$R = R_{Ln} \tag{5-20}$$
$$Z_o = n^{\frac{1}{2}}R \tag{5-21}$$
$$L = jZ_o \tag{5-22}$$
$$C = -jZ_o \tag{5-23}$$
n = number of output ports desired

Note: Port-to-port isolation is a function of precision. An isolation of approximately 30 dB is obtainable with 1 percent accuracy.

WILKINSON POWER DIVIDER

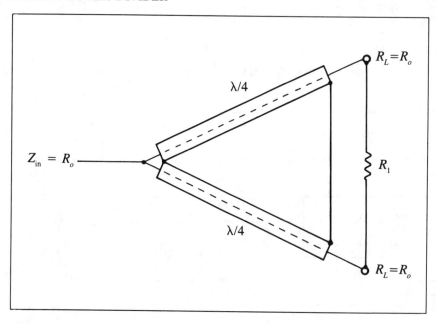

Coaxial Version (where f is 300 to 1000 MHz)

Coaxial Impedance (Z_c)

$$Z_c = 1.414 R_o \tag{5-24}$$

$$\frac{\lambda}{4} = \frac{2951 V_p}{f} \quad \text{(in } inches\text{)} \tag{5-25}$$

$$\frac{\lambda}{4} = \frac{7495 V_p}{f} \quad \text{(in } cm\text{)} \tag{5-26}$$

R_1 is equal $2R_o$
V_p is the velocity of propagation in the cable (see Table 5-1).
f is frequency (in MHz)

Table 5-1. Velocity of Propagation

Medium	V_p
Solid Polyethylene	0.660
Foam Polyethylene	0.780
Teflon	0.695

Lumped Constant Version (where f is less than 300 MHz)

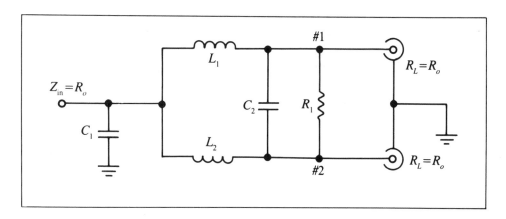

$$C_1 = \frac{1}{2\pi f R_o} \tag{5-27}$$

$$C_2 = \tfrac{1}{2} C_1 \tag{5-28}$$

$$L_1 = L_2 = R_o/(2\pi f) \tag{5-29}$$

$$R_1 = 2 R_o \tag{5-30}$$

f is frequency (in *MHz*)
C is capacitance (in *microfarads*)
L is inductance (in *microhenries*)

6

magnetic circuits

DEFINITIONS

flux Expressed with the greek lowercase letter *phi* (ϕ). The total number of lines of force in a magnetic circuit (in *maxwells*).

flux density Represented by B for square inches or β for square centimeters. The number of lines of force per unit area.

gauss Flux density expressed in lines per square centimeter where one gauss is one line per square centimeter.

hysteresis A materials memory to previous magnetic excitation that must be overcome to restore that material to an unmagnetized state.

permeability Represented by lowercase *mu* (μ). A comparison of a magnetic materials susceptibility to the creation of a magnetic field as compared to air.

rel Represented by R. The cgs measure of reluctance. A material has a reluctance of one rel when one ampere-turn creates one line of force in it.

reluctance (magnetic) The resistance of a material to the creation of a magnetic field within itself expressed in rels.

residual magnetism or **retentivity** The ability of a material to retain some flux after the magnetizing force is removed.

saturation That limit of flux density that cannot be exceeded by the increase of magnetizing force.

MAGNETIC RELATIONSHIPS

$$\phi = F/R \qquad \text{(6-1)}$$
$$F = \phi R \qquad \text{(6-2)}$$
$$R = F/\phi \qquad \text{(6-3)}$$

$$\mu = B/H \qquad \text{(6-4)}$$
$$\mu = \beta/H \qquad \text{(6-5)}$$
$$H = \beta/\mu \qquad \text{(6-6)}$$
$$H = B/\mu \qquad \text{(6-7)}$$
$$\beta = \mu H \qquad \text{(6-8)}$$
$$B = \mu H \qquad \text{(6-9)}$$

where

ϕ = flux (total magnetic lines)

F = magnetomotive force (mmf)

R = reluctance (rels)

μ = permeability

B = flux density in lines per square inch

β = flux density in lines per square centimeter

H = magnetomotive force (mmf) in ampere turns per inch

H = magnetomotive force (mmf) in gilberts per centimeter, or oersteads

B-H CURVE OR HYSTERESIS LOOP

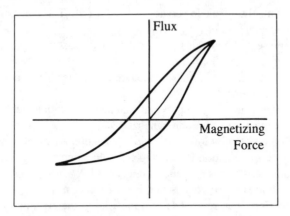

LOSSLESS TRANSFORMERS

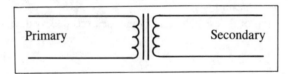

TURNS RATIO (*n*)

$$n = N_s/N_p \tag{6-10}$$

$$n = \sqrt{Z_s/Z_p} \tag{6-11}$$

N_s is the secondary number of turns
N_p is the primary number of turns
Z_s is the secondary impedance
Z_p is the primary impedance

SECONDARY VOLTAGE (*E_s*)

$$E_s = nE_p \tag{6-12}$$

E_p is the primary voltage
n is the turns ratio

IMPEDANCE RATIO

$$Z_{\text{ratio}} = n^2 \tag{6-13}$$

SECONDARY IMPEDANCE (*Z_s*)

$$Z_s = Z_{\text{ratio}}Z_p \tag{6-14}$$

$$Z_s = n^2/Z_p \tag{6-15}$$

Z_p is the primary impedance

PRIMARY IMPEDANCE (*Z_p*)

$$Z_p = Z_s/Z_{\text{ratio}} \tag{6-16}$$

$$Z_p = Z_s/n^2 \tag{6-17}$$

POWER TRANSFORMERS

CORE AREA

$$A_c = \frac{(W/f)^{\frac{1}{2}}}{0.72} \quad (\text{in } in^2) \tag{6-18}$$

PRIMARY TURNS

$$N = 3.49 \cdot 10^6 \left(\frac{E}{A_c B_m f}\right) \tag{6-19}$$

W is transformer output watts
f is frequency (in Hz)
E is the primary voltage (rms)
B_m is flux density (in *gauss*)

AUTO TRANSFORMERS (TOROIDAL) _____

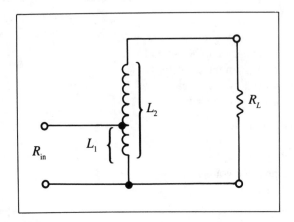

IMPEDANCE RATIO

$$Z_{\text{ratio}} = Z_{L_2}/Z_{L_1} \tag{6-20}$$

$$Z_{\text{ratio}} = R_L/R_{\text{in}} \tag{6-21}$$

TURNS RATIO (*n*)

$$n = \sqrt{Z_{\text{ratio}}} \tag{6-22}$$

TOTAL WINDING (L_2)

$$L_2 = \frac{k \cdot R_L}{\omega} \quad \text{(in } \mu H) \tag{6-23}$$

$k \geq 5$
$\omega = 2\pi f$
f is frequency   (in *MHz*)

TURNS FOR L_2 (N_{total})

$$N_{\text{total}} = 100\sqrt{L_s/L_{100}} \tag{6-24}$$

L_{100} is the inductance per 100 turns

NUMBER OF TURNS FROM THE BOTTOM TO THE TOP (N_{tap})

$$N_{\text{tap}} = N_{\text{total}}/n \tag{6-25}$$

POWDERED IRON OR FERRITE TOROIDAL COILS _____

MAXIMUM FLUX DENSITY (AC EXCITATION ONLY)

$$B_{\text{max}} = \frac{E_{\text{rms}} \, 10^8}{4.44 \, f N A_e} \quad \text{(in } gauss) \tag{6-26}$$

E_{max} is the applied voltage
f is the frequency (in Hz)
N is the number of turns
A_e is the equivalent area of the magnetic path (in cm^2)

When the excitation consists of AC and DC excitation, B_{max} can be found from:

$$B_{max} = \frac{E_{rms}\ 10^8}{4.44\ fNA_e} + \frac{NIA_L}{10\ A_e} \qquad (6\text{-}27)$$

where
I is the DC excitation current
A_L is the core's induction index

INDUCTANCE-TO-TURNS RELATIONSHIP

The Inductance L is related to the number of turns N by

$$N = 100\sqrt{\frac{L}{A_L}} \qquad \begin{array}{l}(\text{in } \mu H) \\ (\text{in } \mu H \text{ per } 100 \text{ turns})\end{array} \qquad (6\text{-}28)$$

or

$$N = 1000\sqrt{\frac{L}{A_L}} \qquad \begin{array}{l}(\text{in } mH) \\ (\text{in } mH \text{ per } 1000 \text{ turns})\end{array} \qquad (6\text{-}29)$$

where
A_L is the core's inductance index

DC-TO-DC CONVERTERS

CORE MATERIALS AND FLUX DENSITY (IN *GAUSS*)

Material	B_m
Orthonol	14,000
48 Alloy	12,000
Hy Mu-80	7,000

FREQUENCY RANGE OF CORE MATERIALS

Tape Thickness (mils)	Frequency (Hz)
1	1600–20000
1 to 2	1000
2 to 4	200
4 to 6	<60

GENERAL EQUATIONS

Turns Ratio (T_r)

$$T_r = \frac{N \text{ output}}{N \text{ primary}} = \frac{E \text{ out}}{E \text{ supply}} \qquad (6\text{-}30)$$

Primary Current (I_p)

$$I_p = T_r I_{\text{Load}} \tag{6-31}$$

Primary Turns (N_p): Faraday's Law

$$N_p = \frac{E}{DB_m AF\ 10^{-8}} \tag{6-32}$$

E is the supply voltage
A is the core cross section (in cm^2)
F is frequency (in Hz)
B_m is flux density
D is 4.44 for sine-wave drive and 4.00 for square-wave drive

Magnetizing Force (H): Ampere's Law

$$H = \frac{0.4\pi NI}{L} \tag{6-33}$$

L is the magnetic path length of the core (in cm)
NI is the ampere-turns
H is in oersteads

Winding Factor (K)

$$K = \frac{NA_\omega}{W} \tag{6-34}$$

W is the core window area (in *circular mils*)
N is turns
A_ω is wire area with insulation (in *circular mils*)

Feedback Turns (N_f)

$$N_f = \frac{N_p}{E} \tag{6-35}$$

E is the supply voltage—use 1 volt of base drive or more
N_p is the primary turns

Feedback Current (I_f)

$$I_f = \frac{I_c}{\beta} \tag{6-36}$$

I_c is collector current
β is h_{fe} or beta of the transistor

HALL EFFECT

HALL OUTPUT VOLTAGE

$$V_H = K_H W(\mathbf{I_d B}) \tag{6-37}$$

K_H is the Hall coefficient (0.1 to 0.3 per ampere-kilogauss typical for an open-circuit measurement)

W is the width of the element

I_d is current density (in *mA*)

B is the magnetic field intensity (in *kilogausses*)

$\mathbf{I_d B}$ is a vector product dependent upon sin θ.

PHYSICAL CONFIGURATION OF HALL EFFECT

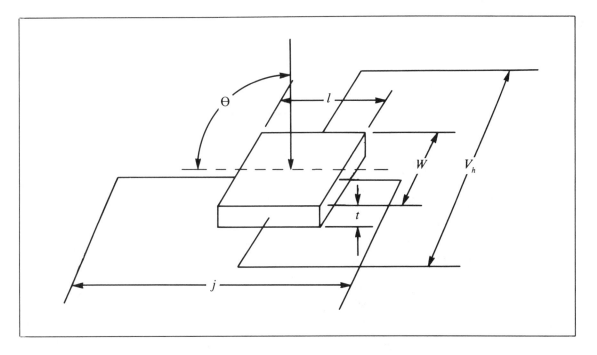

If

$$I_d = \frac{I_c}{Wt} \tag{6-38}$$

where

I_c is the control current

t is the element thickness

Then

$$V_h = K_H \frac{1}{t} (I_c B)$$ (6-39)

or

$$V_H = K_{HOC}(I_c B)$$ (6-40)

K_{HOC} is the open-circuit Hall sensitivity constant

7

---

digital concepts

System	Base or Radix
Binary	2
Ternary	3
Quaternary	4
Quinary	5
Senary	6
Septenary	7
Octanary/Octal	8
Novenary	9
Decimal	10
Duo decimal	12
Sexadecimal/Hexadecimal	16

Format

$$a_1(r)^{n-1} + a_2(r)^{n-2} + a_3(r)^{n-3} + \cdots a_n(r)^{n-n} \tag{7-1}$$

r is the base or radix
a is a digit in the number system
n is the number of digits

EXAMPLE:

If $r = 8$ and $n = 4$, express 7302_8.

$$7(8)^3 + 3(8)^2 + 0(8)^1 + 2(8)^0$$

Table 7-1.   Decimal to Radix 2, 5, 8, or 16

Radix				
10	**2**	**5**	**8**	**16**
0	0	0	0	0
1	1	1	1	1
2	10	2	2	2
3	11	3	3	3
4	100	4	4	4
5	101	10	5	5
6	110	11	6	6
7	111	12	7	7
8	1000	13	10	8
9	1001	14	11	9
10	1010	20	12	A
11	1011	21	13	B
12	1100	22	14	C
13	1101	23	15	D
14	1110	24	16	E
15	1111	30	17	F
16	10000	31	20	10
17	10001	32	21	11
18	10010	33	22	12
19	10011	34	23	13
20	10100	40	24	14
21	10101	41	25	15
22	10110	42	26	16

BINARY NOTATION

Power of Two: 2^n . . . 2^4 2^3 2^2 2^1 2^0
Decimal Value: 16   8   4   2   1

When converting decimal to binary, as in Table 7-2, note that the binary value is selected by the indication of a 1 and rejected by a 0.

Table 7-2. Binary Equivalents of Decimal Numbers

Decimal Value	(16) 2^4	(8) 2^3	(4) 2^2	(2) 2^1	(1) 2^0
0	0	0	0	0	0
1	0	0	0	0	1
2	0	0	0	1	0
3	0	0	0	1	1
4	0	0	1	0	0
5	0	0	1	0	1
6	0	0	1	1	0
7	0	0	1	1	1
8	0	1	0	0	0
9	0	1	0	0	1
10	0	1	0	1	0
11	0	1	0	1	1
12	0	1	1	0	0
13	0	1	1	0	1
14	0	1	1	1	0
15	0	1	1	1	1
16	1	0	0	0	0
17	1	0	0	0	1
18	1	0	0	1	0
19	1	0	0	1	1
20	1	0	1	0	0

The maximum value expressed by a group of 1s and 0s is when all possible values equal 1. The maximum in decimal value equals the sum of all of the individual values.

$$1111 = 1 + 2 + 4 + 8 = 15$$

If there are n bits (1s or 0s), the maximum value is $2^n - 1$. In the previous example, there were four 1s, which has a decimal value of $2^4 - 1$ or $16 - 1 = 15$.

Binary Addition

$$
\begin{array}{cccc}
0 & 0 & 1 & 1 \\
+0 & +1 & +0 & +1 \\
\hline
0 & 1 & 1 & 10 \\
\end{array}
$$

The 1 is a carry.

EXAMPLE:

$$
\begin{array}{rll}
0111 = & 1 + 2 + 4 = & 7 \\
\underline{1010} = & 2 + 8 \qquad = & \underline{10} \\
10001 = & 16 + 1 \qquad = & 17
\end{array}
$$

Binary Subtraction

Method 1 (Direct Method)

$$
\begin{array}{rll}
1011 = & 1 + 2 + 8 = & 11 \\
\underline{-0101} = & 1 + 4 \qquad = & \underline{-5} \\
0110 = & 2 + 4 \qquad = & \;\;6
\end{array}
$$

Method 2 (Complementary Method)

Complement the subtrahend by changing 1s to 0s and 0s to 1s, add 1 to the result, then add, rather than subtract, the final numbers.

EXAMPLE:

$$
\begin{array}{cccc}
1011 & 1011 & 1011 & 1011 \\
\underline{-0101} \quad\rightarrow & \underline{1010} \quad\rightarrow & \underline{1011} \quad\rightarrow & \underline{+1011} \\
 & \text{(complement} & \text{(one added} & \cancel{1}0110 \\
 & \text{subtrahend)} & \text{to subtrahend)} &
\end{array}
$$

Binary Multiplication

$$
\begin{array}{l}
0 \cdot 0 = 0 \\
0 \cdot 1 = 0 \\
1 \cdot 1 = 1 \\
1 \cdot 0 = 0
\end{array}
$$

EXAMPLE:

$$
\begin{array}{rl}
1110 = 2 + 4 + 8 = & 14 \\
\underline{100} \qquad\qquad = & \underline{\times\,4} \\
0000 & 56 \\
0000 & \\
\underline{1110} & \\
111000 = 8 + 16 + 32 = 56 &
\end{array}
$$

Binary Division

The method used is like the decimal system but with binary subtraction rules.

EXAMPLE:

Divide 16 by 3.

$$101_2 = 5 + \text{remainder of 1 or 5 1/3}$$

$$11_2 \overline{)10000_2}$$

$$\underline{11}$$

$$100$$

$$\underline{11}$$

$$1 \text{ (remainder)}$$

EXCESS-THREE CODING

Each decimal digit is represented in binary code plus 3.

Decimal	Excess-3
0	0011
1	0100
2	0101
3	0110
4	0111
5	1000
6	1001
7	1010
8	1011
9	1100

Note that this code allows all 10 decimal values to be represented by four binary digits.

BI-QUINARY NOTATION

Seven digits are used. From right to left, the first two digits are the ''bi'' digits with a value of 0 or 5 as noted by 0 = 01 and 5 = 10. The next five digits represent the numbers 0 to 4.

Decimal Value	Bi-Quinary Coding	
	''bi'' part	''quinary'' part
0	01	00001
1	01	00010
2	01	00100
3	01	01000
4	01	10000
5	10	00001
6	10	00010
7	10	00100
8	10	01000
9	10	10000

THE 9'S COMPLEMENT OF EXCESS-THREE NOTATION

The 9's complement of a number is accomplished by inverting all 1s and 0s.

Decimal Value	4	9	2
Excess Three	0111	1100	0101
9's Complement	1000	0011	1010

THE 9'S AND 10'S COMPLEMENT OF A DECIMAL VALUE

Nine's Complement

To get the 9's complement of a decimal number, subtract the number from a number of equal digits that are all nines.

EXAMPLE:

The 9's complement of 2943 is:

$$
\begin{array}{rl}
9999 & \text{Use a 9 for each digit} \\
-2943 & \text{Subtract the number} \\
\hline
7056 & \text{Answer is in 9's complement}
\end{array}
$$

Ten's Complement

To get the 10's complement of a decimal number, subtract it from the closest larger power of ten.

EXAMPLE:

The 10's complement of 2943 is:

$$
\begin{array}{rl}
10000 & \text{First larger power of ten} \\
-2943 & \text{Subtract the number} \\
\hline
7057 & \text{Answer is in 10's complement}
\end{array}
$$

Note that the 10's complement is simply 1 greater than the 9's complement, and vice versa.

THEOREMS AND POSTULATES

Law of Identity

$$A = A$$
$$\overline{A} = \overline{A}$$

Commutative Law

$$A + B = B + A$$
$$AB = BA$$

Associative Law

$$(AB)C = A(BC)$$
$$(A + B) + C = A + (B + C)$$

Idempotent Law

$$AA = A$$
$$A + A = A$$
$$(BA)(BA) = BA$$

Negation Law

$$\overline{\overline{A}} = A$$
$$\overline{\overline{\overline{AB}} + \overline{C}} = \overline{AB} + \overline{C} = \overline{(AB) + C}$$

Complementary Law

$$\overline{A}A = 0$$
$$\overline{A} + A = 1$$

Intersection Law

$$A \cdot 0 = 0$$
$$A \cdot 1 = 1$$

DeMorgan or Dualization Law

$$\overline{A} + \overline{B} = \overline{AB}$$
$$\overline{AB} = \overline{A} + \overline{B}$$

Distributive Law

$$AB + AC = A(B + C)$$
$$(A + B)(A + C) = A + BC$$

Absorption Law

$$A(A + B) = A$$
$$A + AB = A$$

LOGIC

BOOLEAN EXPRESSIONS

Word	Symbol
Not	—
And	·
Or	+

LOGIC DIAGRAMS _____

AND GATE

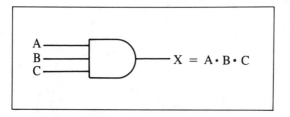

Table 7-3.   AND Truth Table

A	B	C	AND $X = A \cdot B \cdot C$	NAND $\overline{X}$
0	0	0	0	1
0	0	1	0	1
0	1	0	0	1
0	1	1	0	1
1	0	0	0	1
1	0	1	0	1
1	1	0	0	1
1	1	1	1	0

NAND GATE

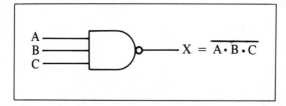

OR GATE

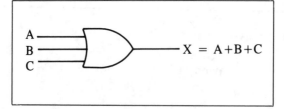

Table 7-4.   OR Truth Table

A	B	C	OR X	NOR $\overline{X}$
0	0	0	0	1
0	0	1	1	0
0	1	0	1	0
0	1	1	1	0
1	0	0	1	0
1	0	1	1	0
1	1	0	1	0
1	1	1	1	0

NOR GATE

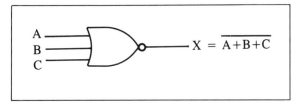

$$X = \overline{A+B+C}$$

EXCLUSIVE OR (EX-OR) AND EX-NOR

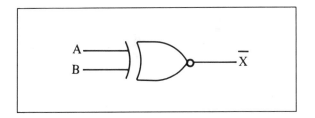

Table 7-5.   EX-OR/EX-NOR
Truth Table

A	B	X	$\overline{X}$
0	1	0	1
0	0	1	0
1	1	1	0
1	0	0	1

HALF-ADDER

A carry occurs only when A and B are both 1s.
A sum occurs only when A or B are 1s with no carry.

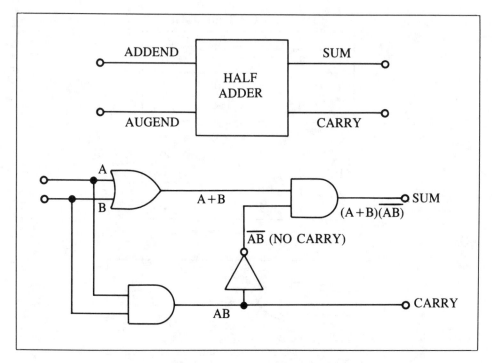

Table 7-6.   Half-Adder Truth Table

A	B	CARRY AB	(A + B)	($\overline{AB}$)	SUM
0	0	0	0	1	0
0	1	0	1	1	1
1	0	0	1	1	1
1	1	1	1	0	0

HALF-SUBTRACTOR

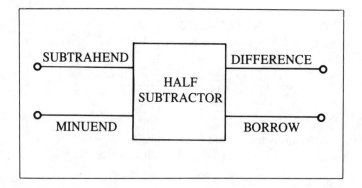

The same configuration as for the half-adder provides the difference output (it is identical to the sum output obtained in the half-adder). Therefore, similar circuitry is used for the subtraction function except that the carry AB is not used and an $\overline{AB}$ must be provided for the borrow.

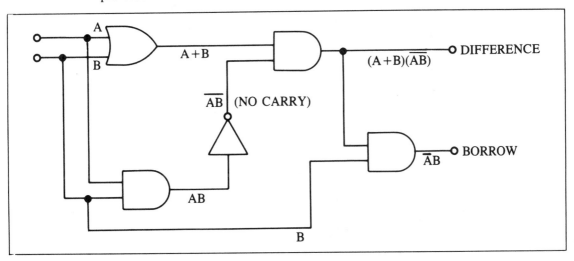

Table 7-7. Half-Subtractor Truth Table

A	B	(A + B)	(AB)	$\overline{(AB)}$	CARRY $(A + B)\,\overline{(AB)}$ (B)	DIFFERENCE $(A + B)\,\overline{(AB)}$
0	0	0	0	1	0	0
0	1	1	0	1	1	1
1	0	1	0	1	0	1
1	1	1	1	0	0	0

PSEUDO-RANDOM SHIFT REGISTERS OF MAXIMUM LENGTH____

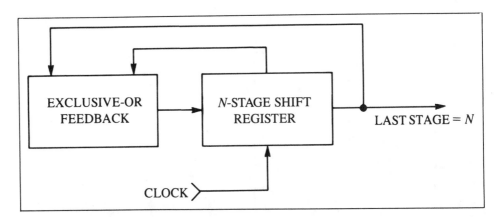

L = Sequence Length
N = Shift Register Stages
F = Feedback From Stage N

L	N	F
3	2	1
7	3	1
7	3	2
15	4	1
15	4	3
31	5	2
31	5	3
63	6	1
63	6	5
127	7	1
127	7	3
127	7	4
127	7	6
511	9	4
511	9	5
1023	10	3
1023	10	7
2047	11	2
2047	11	9
32767	15	1
32767	15	4
32767	15	7
32767	15	8
32767	15	11
32767	15	14
131071	17	3
131071	17	5
131071	17	12
131071	17	14
262143	18	7
262143	18	11
1048575	20	3
1048575	20	17
2097151	21	2
2097151	21	19
4194303	22	1

Table 7-8.   Shift Register Feedback for Maximum-Length Sequences

L	N	F
4194303	22	21
8388607	23	5
8388607	23	9
8388607	23	14
8388607	23	18
33554431	25	3
33554431	25	7
33554431	25	18
33554431	25	22
133693177	27	8
133693177	27	19
268435455	28	3
268435455	28	9
268435455	28	13
268435455	28	15
268435455	28	19
268435455	28	25
536870911	29	2
536870911	29	27
2147483647	31	3
2147483647	31	6
2147483647	31	7
2147483647	31	13
2147483647	31	18
2147483647	31	24
2147483647	31	25

TRANSISTOR-TRANSISTOR LOGIC (TTL)

NUMBER CODE

The TTL number code consists of the device identification, the package identification letter, and the temperature range identification letter.

Package

D   Dual in-line package (DIP), ceramic, hermetic
E   Plastic can
F   Flat pack, hermetic
H   Metal can
P   Plastic dual in-line package (DIP).

Temperature

C   Commercial, 0 to 70 or 75 C, (7400 series)
M   Military, −55 to 125 C, (5400 series)

Table 7-9.   TTL High and Low Potentials

Condition	Specified	Typical
V_{oh} = High state Output volts (min)	2.0 V	2.4 to 2.7 V
V_{ol} = Low state Output volts (max)	0.8* V	0.3 to 0.5 V
*0.7 for "L" Series		

TTL LOADING

Unit Load (UL)

Logic 1 = 40 μA
Logic 0 = 1.6 mA

An N-unit Load

Logic 1 = N 40 μA
Logic 0 = N 1.6 mA

TTL Series	UNIT LOADS			
	Input		Output	
	1	0	1	0
5400, 7400	1	1	20	10
54H00, 74H00	1.25	1.25	25	12.5
54S00, 74S00	1.25	1.25	25	12.5
54L00, 74L00	.5	.25	10	2.5

PULL-UP RESISTORS FOR TTL

$$R_{\min} = \frac{V_{cc\ max} - 0.4}{I_f - N_i\ 1.6(ma)}$$

(7-1)

$$R_{\max} = \frac{V_{cc\ \min} - 2.4}{N_o\ I_{hl} + N_i\ 40(\mu A)}$$

(7-2)

I_f = LOW fan-out current of driver
N_i = number of driven input loads
N_o = number of wired OR outputs
I_{hl} = HIGH output leakage current
V_{cc} is the supply potential

8

transistors

TRANSISTOR PARAMETERS

Z PARAMETERS (OPEN CIRCUIT)

Condition: frequency must be less than 2 MHz.

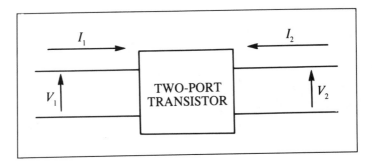

Input Impedance

$$Z_{11} = Z_i = V_1/I_1, \ I_2 = 0 \qquad (8\text{-}1)$$

Forward Transfer Impedance

$$Z_{21} = Z_f = V_2/I_1, \ I_2 = 0 \qquad (8\text{-}2)$$

Output Impedance

$$Z_{22} = Z_o = V_2/I_2, \ I_1 = 0 \qquad (8\text{-}3)$$

Reverse Transfer Impedance

$$Z_{12} = Z_r = V_1/I_2, \; I_1 = 0 \tag{8-4}$$

$$V_1 = Z_i I_1 + Z_r I_2 \tag{8-5}$$
$$V_2 = Z_f I_1 + Z_o I_2 \tag{8-6}$$

$$\begin{bmatrix} V_1 \\ V_2 \end{bmatrix} = \begin{bmatrix} Z_i & Z_r \\ Z_f & Z_o \end{bmatrix} \begin{bmatrix} I_1 \\ I_2 \end{bmatrix}$$

y PARAMETERS (SHORT CIRCUIT)

Condition: frequency less than 200 MHz.

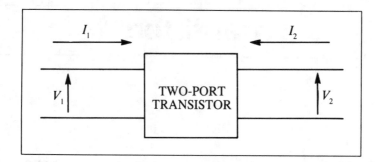

Input Admittance

$$y_{11} = y_i = I_1/V_1, \; V_2 = 0 \tag{8-7}$$

Forward Transmission Admittance

$$y_{21} = y_f = I_2/V_1, \; V_2 = 0 \tag{8-8}$$

Output Admittance

$$y_{22} = y_o = I_2/V_2, \; V_1 = 0 \tag{8-9}$$

Reverse Transmission Admittance

$$y_{12} = y_r = I_1/V_2, \; V_1 = 0 \tag{8-10}$$

$$I_1 = y_i V_1 + y_r V_2 \tag{8-11}$$
$$I_2 = y_f V_1 + y_o V_2 \tag{8-12}$$

$$\begin{bmatrix} I_1 \\ I_2 \end{bmatrix} = \begin{bmatrix} y_i & y_r \\ y_f & y_o \end{bmatrix} \begin{bmatrix} V_1 \\ V_2 \end{bmatrix}$$

CONVERTING ADMITTANCE PARAMETER DATA BETWEEN CB, CE, AND CC TRANSISTOR CONFIGURATIONS

Below is a sample table for doing admittance conversion. All data must first be converted to admittance format.

	Base	Emitter	Collector
Base	—	—	—
Emitter	—	—	—
Collector	—	—	—

Follow these steps to fill in the matrix.

• Given *y*-parameter data for any configuration, for example common base, cross out that row and column, as below.

	Base	Emitter	Collector
Base	—┼—	═══	═══
Emitter	—┼—	y_{ib}	y_{rb}
Collector	—┼—	y_{fb}	y_{ob}

• Enter data into the four remaining areas in matrix format as follows. The sum of any row and column must equal zero. Fill in the four blank spaces to satisfy this rule. This completes the universal matrix.

• To convert to any other configuration, such as common collector, cross out the collector row and column. The data that remains is for the common-collector configuration.

	Base	Emitter	Collector
Base	y_{ic}	y_{rc}	—┼—
Emitter	y_{fc}	y_{oc}	—┼—
Collector	═══	═══	═┼═

h PARAMETERS (HYBRID)

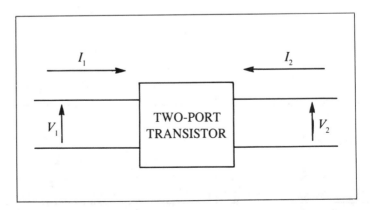

Input Impedance

$$h_{11} = h_i = V_1/I_1, \ V_2 = 0 \qquad (8\text{-}13)$$

Forward Transfer Current Ratio

$$h_{21} = h_f = I_2/I_1, \ V_2 = 0 \qquad (8\text{-}14)$$

Output Admittance

$$h_{22} = h_o = I_2/V_2, \ I_1 = 0 \qquad (8\text{-}15)$$

Reverse Transfer Voltage Ratio

$$h_{12} = h_r = V_1/V_2, \ I_1 = 0 \qquad (8\text{-}16)$$

$$V_1 = h_i I_1 + h_r V_2 \qquad (8\text{-}17)$$
$$I_2 = h_f I_1 + h_o V_2 \qquad (8\text{-}18)$$

$$\begin{bmatrix} V_1 \\ I_2 \end{bmatrix} = \begin{bmatrix} h_i & h_r \\ h_f & h_o \end{bmatrix} \begin{bmatrix} I_1 \\ V_2 \end{bmatrix}$$

Table 8-1. h,y,z Conversion

From	h		y		z	
To						
h	h_i	h_r	$\dfrac{1}{y_i}$	$\dfrac{-y_r}{y_i}$	$\dfrac{D_z}{z_o}$	$\dfrac{z_r}{z_o}$
	h_f	h_o	$\dfrac{y_f}{y_i}$	$\dfrac{D_y}{y_i}$	$\dfrac{-z_f}{z_o}$	$\dfrac{1}{z_o}$
y	$\dfrac{1}{h_i}$	$\dfrac{-h_r}{h_i}$	y_i	y_r	$\dfrac{z_o}{D_z}$	$\dfrac{-z_r}{D_z}$
	$\dfrac{h_f}{h_i}$	$\dfrac{D_h}{h_i}$	y_f	y_o	$\dfrac{-z_f}{D_z}$	$\dfrac{z_i}{D_z}$
z	$\dfrac{D_h}{h_o}$	$\dfrac{h_r}{h_o}$	$\dfrac{y_o}{D_y}$	$\dfrac{-y_r}{D_y}$	z_i	z_r
	$\dfrac{-h_f}{h_o}$	$\dfrac{1}{h_o}$	$\dfrac{-y_f}{D_y}$	$\dfrac{y_i}{D_y}$	z_f	z_o

D is the determinant of the parameters. For example,

$$\begin{bmatrix} h_i & h_r \\ h_f & h_o \end{bmatrix}$$

$$D_h = h_i\,h_o - h_f\,h_r \tag{8-19}$$

TRANSISTOR AMPLIFIERS

h Parameters for Small-Signal Amplifiers

Maximum Available Power Gain (MAG)

$$\text{MAG} = \frac{h_f{}^2}{h_i\,h_o\left[1 + \left(1 - \dfrac{h_f\,h_r}{h_i\,h_o}\right)\right]^2} \tag{8-20}$$

Optimum Load Resistance for Match

$$R_{L_{\text{opt}}} = \frac{1}{h_o\left(1 - \dfrac{h_f\,h_r}{h_i\,h_o}\right)^{\frac{1}{2}}} \tag{8-21}$$

Optimum Source Resistance for Match

$$R_{S_{\text{opt}}} = h_i\left(1 - \frac{h_f\,h_r}{h_i\,h_o}\right)^{\frac{1}{2}} \tag{8-22}$$

Power Gain for Any Source and Load Resistance

$$P_g = \frac{4\,R_S}{R_L}\left[\frac{h_f}{\left(h_i + R_S\right)\left(h_o + \dfrac{1}{R_L}\right) - h_f\,h_r}\right]^2 \tag{8-23}$$

Gain of Low-Frequency Power Amplifiers

Common Emitter

$$P_g = h_{fe}{}^2 R_L / h_{ie} \tag{8-24}$$

Common Base

$$P_g = h_{fe}\,R_L / h_{ie} \tag{8-25}$$

Common Collector

$$P_g = h_{fe} \qquad \text{(where } R_L \ll Z_o\text{)} \tag{8-26}$$

Load Resistance Maximum

$$R_{L_{\text{max}}} = \frac{V_{cc}{}^2}{2\,P_{o_{\text{max}}}} \tag{8-27}$$

$P_{o_{max}}$ is the maximum usable output.

THREE-PORT SCATTERING PARAMETERS (S)

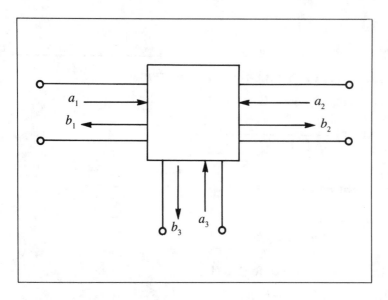

INPUT REFLECTION COEFFICIENT—PORTS 2 AND 3 MATCHED

$$S_{11} = \frac{b_1}{a_1} \quad a_2, a_3 = 0$$

FORWARD TRANSMISSION COEFFICIENT (1 TO 2)—PORTS 2 AND 3 Matched

$$S_{21} = \frac{b_2}{a_1} \quad a_2, a_3 = 0$$

FORWARD TRANSMISSION COEFFICIENT (1 TO 3)—PORTS 2 AND 3 Matched

$$S_{31} = \frac{b_3}{a_1} \quad a_2, a_3 = 0$$

These parameters continue to follow the pattern established.

$$\begin{bmatrix} b_1 \\ b_2 \\ b_3 \end{bmatrix} = \begin{bmatrix} S_{11} & S_{12} & S_{13} \\ S_{21} & S_{22} & S_{23} \\ S_{31} & S_{32} & S_{33} \end{bmatrix} \begin{bmatrix} a_1 \\ a_2 \\ a_3 \end{bmatrix}$$

TWO-PORT SCATTERING PARAMETERS (S)

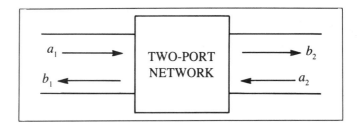

INPUT REFLECTION COEFFICIENT—MATCHED OUTPUT

$$S_{11} = \frac{b_1}{a_1} \qquad a_2 = 0$$

FORWARD TRANSMISSION COEFFICIENT—MATCHED OUTPUT

$$S_{21} = \frac{b_2}{a_1} \qquad a_2 = 0$$

OUTPUT REFLECTION COEFFICIENT—MATCHED INPUT

$$S_{22} = \frac{b_2}{a_2} \qquad a_1 = 0$$

REVERSE TRANSMISSION COEFFICIENT—MATCHED INPUT

$$S_{12} = \frac{b_1}{a_2} \qquad a_1 = 0$$

$$b_1 = S_{11}\, a_1 + S_{12}\, a_2 \qquad\qquad (8\text{-}28)$$
$$b_2 = S_{21}\, a_1 + S_{22}\, a_2 \qquad\qquad (8\text{-}29)$$

$$\begin{bmatrix} b_1 \\ b_2 \end{bmatrix} = \begin{bmatrix} S_{11} & S_{12} \\ S_{21} & S_{22} \end{bmatrix} \begin{bmatrix} a_1 \\ a_2 \end{bmatrix}$$

CONVERSION EQUATIONS NORMALIZED TO Z_O

Actual Parameters	Normalized Value Relative to Actual Value
$z_{ii}{}'$	$z_{ii} \cdot Z_o$
$y_{ii}{}'$	y_{ii}/Z_o
$h_{11}{}'$	$h_{11} \cdot Z_o$
$h_{12}{}'$	h_{12}
$h_{21}{}'$	h_{21}
$h_{22}{}'$	h_{22}/Z_o

Actual values are shown primed. Normalized values to Z_o are unprimed.

z TO S

$$S_{11} = \frac{(z_{11} - 1)(z_{22} + 1) - z_{12} z_{21}}{\Delta z} \qquad (8\text{-}30)$$

$$S_{12} = \frac{2z_{12}}{\Delta z} \qquad (8\text{-}31)$$

$$S_{21} = \frac{2z_{21}}{\Delta z} \qquad (8\text{-}32)$$

$$S_{22} = \frac{(z_{11} + 1)(z_{22} - 1) - z_{12} z_{21}}{\Delta z} \qquad (8\text{-}33)$$

$$\Delta z = (z_{11} + 1)(z_{22} + 1) - z_{12} z_{21} \qquad (8\text{-}34)$$

y TO S

$$S_{11} = \frac{(1 - y_{11})(1 + y_{22}) + y_{12} y_{21}}{\Delta y} \qquad (8\text{-}35)$$

$$S_{12} = \frac{-2y_{12}}{\Delta y} \qquad (8\text{-}36)$$

$$S_{21} = \frac{-2y_{21}}{\Delta y} \qquad (8\text{-}37)$$

$$S_{22} = \frac{(1 + y_{11})(1 - y_{22}) + y_{12} y_{21}}{\Delta y} \qquad (8\text{-}38)$$

$$\Delta y = (1 + y_{11})(1 + y_{22}) - y_{12} y_{21} \qquad (8\text{-}39)$$

h TO S

$$S_{11} = \frac{(h_{11} - 1)(1 + h_{22}) - h_{12} h_{21}}{\Delta h} \qquad (8\text{-}40)$$

$$S_{12} = \frac{2h_{12}}{\Delta h} \qquad (8\text{-}41)$$

$$S_{21} = \frac{-2h_{21}}{\Delta h} \qquad (8\text{-}42)$$

$$S_{22} = \frac{(1 + h_{11})(1 - h_{22}) + h_{12} h_{21}}{\Delta h} \qquad (8\text{-}43)$$

$$\Delta h = (1 + h_{11})(1 + h_{22}) - h_{12} h_{21} \qquad (8\text{-}44)$$

S TO z

$$z_{11} = \frac{(1 + S_{11})(1 - S_{22}) + S_{12} S_{21}}{\Delta sz} \qquad (8\text{-}45)$$

$$z_{12} = \frac{2S_{12}}{\Delta sz} \tag{8-46}$$

$$z_{21} = \frac{2S_{21}}{\Delta sz} \tag{8-47}$$

$$z_{22} = \frac{(1 - S_{11})(1 + S_{22}) + S_{12} S_{21}}{\Delta sz} \tag{8-48}$$

$$\Delta sz = (1 - S_{11})(1 - S_{22}) - S_{12} S_{21} \tag{8-49}$$

S TO y

$$y_{11} = \frac{(1 - S_{11})(1 + S_{22}) + S_{12} S_{21}}{\Delta sy} \tag{8-50}$$

$$y_{12} = \frac{-2S_{12}}{\Delta sy} \tag{8-51}$$

$$y_{21} = \frac{-2S_{21}}{\Delta sy} \tag{8-52}$$

$$y_{22} = \frac{(1 + S_{11})(1 - S_{22}) + S_{12} S_{21}}{\Delta sy} \tag{8-53}$$

$$\Delta sy = (1 + S_{11})(1 + S_{22}) - S_{12} S_{21} \tag{8-54}$$

S TO h

$$h_{11} = \frac{(1 + S_{11})(1 + S_{22}) - S_{12} S_{21}}{\Delta sh} \tag{8-55}$$

$$h_{12} = \frac{2S_{12}}{\Delta sh} \tag{8-56}$$

$$h_{21} = \frac{-2S_{21}}{\Delta sh} \tag{8-57}$$

$$h_{22} = \frac{(1 - S_{11})(1 - S_{22}) - S_{12} S_{21}}{\Delta sh} \tag{8-58}$$

$$\Delta sh = (1 - S_{11})(1 + S_{22}) + S_{12} S_{21} \tag{8-59}$$

S-PARAMETER RELATIONSHIPS

S_{11} FROM Z_{in} AND Z_{out}

$$S_{11} = \frac{Z_{in} - Z_o}{Z_{in} + Z_o} \tag{8-60}$$

S_{22} FROM Z_{in} AND Z_{out}

$$S_{22} = \frac{Z_{out} - Z_o}{Z_{out} + Z_o} \tag{8-61}$$

where Z_o is the termination impedance, usually 50 ohms.

FORWARD GAIN AND REVERSE GAIN

$$\text{Forward Gain} = |S_{21}|^2 \tag{8-62}$$

$$\text{Reverse Gain} = |S_{12}|^2 \tag{8-63}$$

Z_{in} FROM Z_o AND S_{11}

$$Z_{in} = Z_o \left[\frac{1 + S_{11}}{1 - S_{11}} \right] \tag{8-64}$$

Z_{out} FROM Z_o AND S_{22}

$$Z_{out} = Z_o \left[\frac{1 + S_{22}}{1 - S_{22}} \right] \tag{8-65}$$

SCATTERING PARAMETERS S_{11}' AND S_{22}' (NONMATCHED CONDITIONS)

$$Z_L \neq Z_o$$

$$S_{11}' = S_{11} + \frac{S_{12} S_{21} \Gamma_L}{1 - S_{22} \Gamma_L} \tag{8-66}$$

$$\Gamma_L = \frac{Z_L - Z_o}{Z_L + Z_o}$$

Note: Unprimed values are matched values.

$$S_{22}' = S_{22} + \frac{S_{12} S_{21} \Gamma_s}{1 - S_{11} \Gamma_s} \tag{8-68}$$

$$\Gamma_s = \frac{Z_g - Z_o}{Z_g + Z_o} \tag{8-69}$$

$$Z_{in} = Z_o \left[\frac{1 + S_{11}'}{1 - S_{11}'} \right] \tag{8-70}$$

$$Z_{out} = Z_o \left[\frac{1 + S_{22}'}{1 - S_{22}'} \right] \tag{8-71}$$

where

Z_L is the load impedance
Z_g is the generator impedance

UNIJUNCTION TRANSISTOR

The unijunction transistor is a negative resistance device. The input impedance when not conducting is typically 5 MΩ.

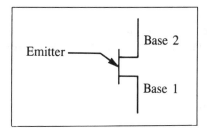

CHARACTERISTIC CURVE

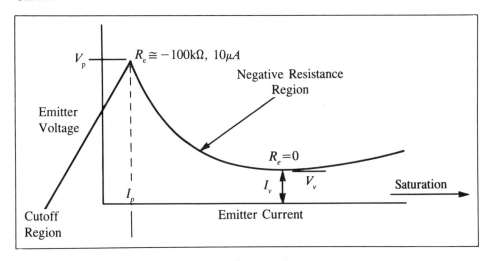

V_p is V_{peak}
I_p is I_{peak}
V_v is V_{valley}
I_v is I_{valley}

EQUIVALENT CIRCUIT (APPROXIMATE)

The figure depicts the off condition.

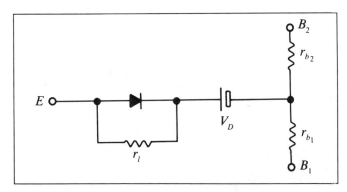

r_l is leakage resistance

V_D is the diode contact potential

r_{B_1} and r_{B_2} are the base-1 and base-2 resistances respectively

The point of maximum negative resistance occurs at V_p, I_p, and the peak voltage to bias the transistor into negative resistance is

$$V_p = \eta\, V_{bb} + V_D \tag{8-72}$$

where

$$\eta = r_{b_1}/(r_{b_1} + r_{b_2}) \tag{8-73}$$

V_{bb} is the supply voltage

MULTIVIBRATORS

Astable Multivibrator (Free-running and Triangular Types)

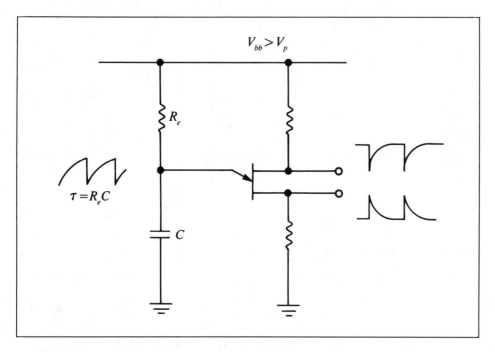

Refer back to the characteristic curve. The emitter load line must intersect the curve in the negative resistance region. The emitter current must be less than I_v at V_v (subscript v represents *valley*).

The frequency of oscillation is

$$f = \cfrac{1}{R_eC\left(\cfrac{1}{1-\eta}\right)} \tag{8-74}$$

Astable Multivibrator (Rectangular)

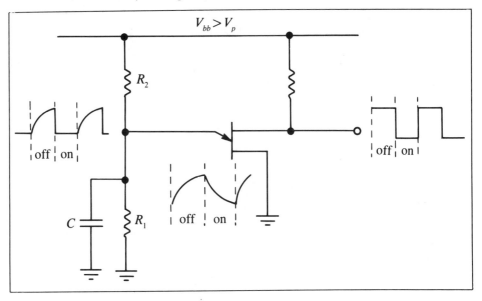

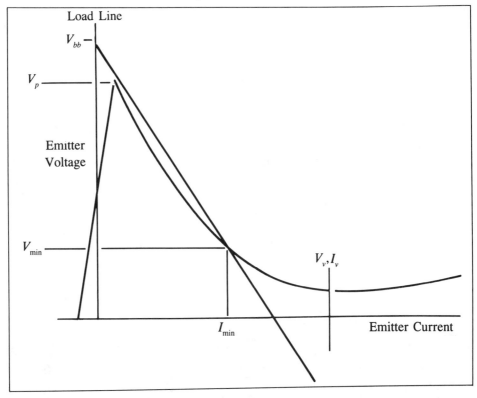

The emitter load line (R_2) must intersect the characteristic curve below V_{valley} at V_{min}, I_{min}.

$$t_{on} = R_1 C \ln V_p / I_{min} \tag{8-75}$$

$$t_{off} = \left(\frac{R_1 R_2}{R_1 + R_2} \right) C \ln \left[\frac{1}{1 - \eta \left(\dfrac{R_1 + R_2}{R_1} \right)} \right] \tag{8-76}$$

where $V_{min} \ll V_{bb}$

If V_{min} is not very much less than V_{bb}, then

$$t_{off} = \left(\frac{R_1 R_2}{R_1 + R_2} \right) C \ln \left(\frac{\dfrac{V_{bb} R_1}{R_1 + R_2} - V_{min}}{V_{bb} \dfrac{R_1}{R_1 + R_2} - V_p} \right) \tag{8-77}$$

9

---

diodes

Pin diodes have a variable resistance characteristic related to the applied bias current.

TYPICAL PERFORMANCE

Bias Current	Resistance
1 μA	10,000 Ω
100 mA	1 Ω

CHARACTERISTICS (EXCLUDING END POINTS)

$$R = R_{1\text{ mA}}I^{-x} \quad \text{(in } ohms\text{)} \tag{9-1}$$

$R_{1\text{ mA}}$ is the resistance of the diode at a bias current of 1 mA
I is the forward bias current
x is the slope of the R-I characteristic

LOW-FREQUENCY OPERATIONAL LIMIT (f_m)

The diode is resistive above this limit.

$$f_m = 1.57\,\frac{1}{\tau} \tag{9-2}$$

τ is the diode lifetime (0.1 to 2.0 μs typical)

Low-frequency operation requires the use of diodes with long lifetimes.

ATTENUATOR DESIGN USING PIN DIODES

DC blocking and RF coupling capacitors are not shown.

T Type

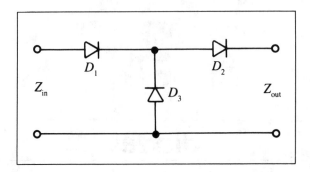

$$Z_{in} = Z_{out} \tag{9-3}$$

$$D_1 = D_2 = Z_{out}(k - 1)/(k + 1) \tag{9-4}$$

$$D_3 = 2 Z_{out}/[k - (1/k)] \tag{9-5}$$

Bridge T

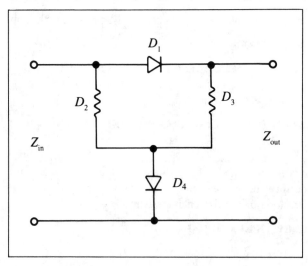

$$Z_{in} = Z_{out} \tag{9-6}$$

$$D_1 = Z_{out}(k - 1) \tag{9-7}$$

$$D_2 = D_3 = Z_{out} \tag{9-8}$$

$$D_4 = Z_{out}\left(\frac{1}{k - 1}\right) \tag{9-9}$$

Pi Type

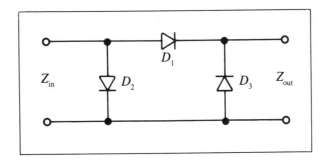

$$Z_{in} = Z_{out} \tag{9-10}$$

$$D_1 = \frac{1}{2}Z_{out}\left(k - \frac{1}{k}\right) \tag{9-11}$$

$$D_2 = D_3 = Z_{out}\left(\frac{k + 1}{k - 1}\right) \tag{9-12}$$

$$k = \frac{\text{input voltage}}{\text{output voltage}}$$

The design procedure is as follows:

• Plot resistances verses k.
• Design the drive circuitry for each element with coupling and isolation components.

TUNING DIODES (VOLTAGE-VARIABLE CAPACITANCE)

EQUIVALENT CIRCUIT

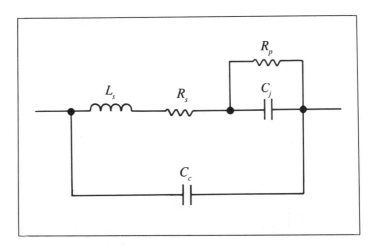

R_s is series resistance
C_c is case capacitance
R_p is junction parallel resistance (usually negligible)
L_s is lead inductance
C_j is the voltage-variable capacitance

CAPACITANCE AT ANY VOLTAGE V

$$C = C_0 \left[\frac{1}{(1 + V/\theta)^\gamma} + \frac{\theta^\gamma}{(\theta + V)^\gamma} \right]$$ (9-13)

C_0 is the capacitance at 0 volts bias
V is the diode reverse-bias voltage
θ is the contact potential
γ is the power law of the junction ($\frac{1}{2}$ for step junctions)

The range of usefulness is from audio frequencies to 2000 MHz.

ADMITTANCE (REFERENCE EQUIVALENT CIRCUIT)

$$y = j\omega C_c + \frac{1}{(R_s + j\omega L_s) + 1/[(1/R_p) + (j\omega C_j)]}$$ (9-14)

Note that R_p can be omitted if sufficiently large.

QUALITY FACTOR OF THE JUNCTION

$$Q_j = 1/(\omega R_s C_j)$$ (9-15)

$\omega = 2\pi f$
f is frequency (in Hz)

CAPACITANCE—GENERAL CASE

$$C = C_c + \frac{C_j}{1 + \omega^2 L_s C_j}$$ (9-16)

$$Q_j > (1 - \omega^2 L_s C_j)$$ (9-17)

DIODE RESONANT FREQUENCY

$$f_d = 1/(2\pi L_s C_j)$$ (9-18)

CAPACITANCE BELOW RESONANCE

Capacitance when $f < f_d$ (f_d is defined following equation (9-18)) is

$$C = C_c + C_j$$ (9-19)

In terms of equation (9-13):

$$C = C_c + C_0 \left[\frac{1}{(1 + V/\theta)^\gamma} + \frac{\theta^\gamma}{(\theta + V)^\gamma} \right]$$ (9-20)

$$C \approx C_c + C_0 \frac{1}{(1 + V/\theta)^\gamma} \tag{9-21}$$

For the step junction:

$$C \approx C_c + C_o \frac{1}{(1 + 2V)^{\frac{1}{2}}} \tag{9-22}$$

QUALITY FACTOR OF THE DIODE

$$Q_d = \frac{1}{\omega R_s C} \tag{9-23}$$

HOT-CARRIER DIODES AND SCHOTTKY DIODES

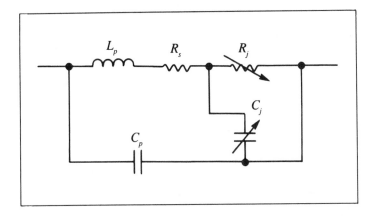

ELECTRICAL CHARACTERISTICS

$$i = I_s \left[\left(\exp \frac{qv}{nkT} \right) - 1 \right] \tag{9-24}$$

I_s is the saturation current
q is the electron charge (1.6×10^{-19} coulombs)
v is the diode junction voltage
n is the ideality factor of the diode
k is Boltzmann's constant (1.38×10^{-23} joules/°K)

At room temperature (300°K), $n \approx 1$. Then,

$$i = I_s\{[\exp(v/26)] - 1\} \quad \text{(typical)} \tag{9-25}$$

R_j(junction resistance) = $(26/I_s) \exp(-v/26)$

C_j(junction capacitance) = $C_{jo}/(1 - v/V_b)^{\frac{1}{2}}$

  C_{jo} is the zero bias capacity of the junction

  V_b is ≈ 0.45 volts

R_s is the series resistance

C_p is packaging capacitance

L_p is package inductance

HOT-CARRIER DIODE RF BANDWIDTH CUTOFF FREQUENCY

$$f_c = \sqrt{1 + (R_s/R_j)} \cdot \frac{1}{2\pi C_j \sqrt{R_s R_j}}$$ (9-26)

TUNNEL DIODES

SYMBOLS

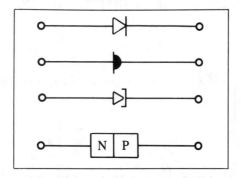

STATIC CHARACTERISTICS

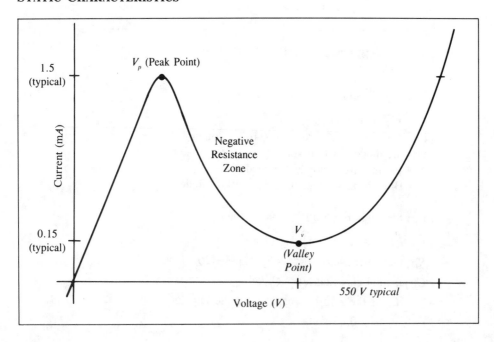

EQUIVALENT CIRCUIT

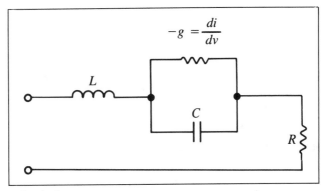

L is series inductance
R is series resistance
$-g$ is negative conductance
C is capacitance

SELF-RESONANT FREQUENCY

$$f = \frac{1}{2\pi}\left[\frac{1}{LC} - \left(\frac{g}{C}\right)^2\right]^{\frac{1}{2}}$$ (9-27)

RESISTIVE CUTOFF FREQUENCY

$$f_r = \left|\frac{g}{2\pi C}\right|\left[\frac{1}{R|g|} - 1\right]^{\frac{1}{2}}$$ (9-28)

AMPLIFIERS

The following assumes that L and R are small.

Equivalent Circuit

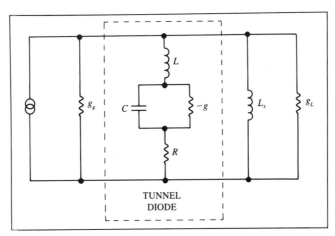

Resonance

At resonance, $L_x > L$.

$$f_o = \frac{1}{2\pi(L_xC)^{\frac{1}{2}}}$$ (9-29)

Power Gain

$$A = 4\frac{g_gg_L}{(g_g + g_L - g)^2}$$ (9-30)

Power Gain at Other Than Resonance

$$A = 4\frac{g_gg_L}{(g_g + g_L - g)^2 + \omega^2C^2\left(1 - \frac{\omega_0^2}{\omega^2}\right)^2}$$ (9-31)

Bandwidth

$$BW = (g_g + g_L - g)\frac{1}{2\pi C}$$ (9-32)

Stability

$$g_g + g_L - g \quad \text{(must be positive)}$$ (9-33)

OSCILLATORS

$$g_g + g_L - g \quad \text{(must be negative)}$$ (9-34)

10

---

amplifiers

STABILITY

Rollett's Stability Factor (K)

For unconditional stability, $K > 1$. Potential instability results when $K < 1$.

$$K = \frac{1 + |S_{11}S_{22} - S_{12}S_{21}|^2 - |S_{11}|^2 - |S_{22}|^2}{2|S_{12}||S_{21}|} \tag{10-1}$$

Linvill C Factor

$$C = K^{-1} \tag{10-2}$$

Absolute Stability

If the following conditions are met, no passive load or source can cause instability.

$$|S_{11}| < 1 \tag{10-3}$$

$$|S_{22}| < 1 \tag{10-4}$$

$$\left| \frac{|S_{12}S_{21}| - |(S_{11} - \Delta S_{22}*)*|}{|S_{11}|^2 - |\Delta|^2} \right| > 1 \tag{10-5}$$

$$\left| \frac{|S_{12}S_{21}| - |(S_{22} - \Delta S_{11}*)*|}{|S_{22}|^2 - |\Delta|^2} \right| > 1 \tag{10-6}$$

where

$$\Delta = S_{11}S_{22} - S_{12}S_{21} \qquad (10\text{-}7)$$

* Denotes the complex conjugate.

Unconditional Stability

$K > 1$ and positive. The conjugate match input reflection coefficient is

$$R_{ms} = C_1{}^* \left[\frac{B_1 \pm \sqrt{B_1{}^2 - 4|C_1|^2}}{2|C_1|^2} \right] \qquad (10\text{-}8)$$

$$C_1 = S_{11} - \Delta S_{22}{}^* \qquad (10\text{-}9)$$
$$B_1 = 1 + |S_{11}|^2 - |S_{22}|^2 - |\Delta|^2 \qquad (10\text{-}10)$$
$$\Delta = S_{11}S_{22} - S_{12}S_{21} \qquad (10\text{-}11)$$

If B_n is negative, the sign before the radical is positive and vice versa (for $n = 1$ to 2).

The conjugate match output reflection coefficient is

$$R_{ml} = C_2{}^* \left[\frac{B_2 \pm \sqrt{B_2{}^2 - 4|C_2|^2}}{2|C_2|^2} \right] \qquad (10\text{-}12)$$

$$C_2 = S_{22} - \Delta S_{11}{}^* \qquad (10\text{-}13)$$
$$B_2 = 1 + |S_{22}|^2 - |S_{11}|^2 - |\Delta|^2 \qquad (10\text{-}14)$$
$$\Delta = S_{11}S_{22} - S_{12}S_{21} \qquad (10\text{-}15)$$

Again, if B_n is negative, the sign before the radical is positive and vice versa.

* Denotes the complex conjugate.

Stability Circles (Unstable within the Circles)

Input Plane: Location of the center of the circle from the center of the Smith chart.

$$r_{s1} = \frac{C_1{}^*}{|S_{11}|^2 - |\Delta|^2} \qquad (10\text{-}16)$$

Radius of the circle located at r_{s1} is

$$R_{s1} = \frac{|S_{12}S_{21}|}{|S_{11}|^2 - |\Delta|^2} \qquad (10\text{-}17)$$

Output Plane: Location of the circle from the center of the Smith chart.

$$r_{s2} = \frac{C_2{}^*}{|S_{22}|^2 - |\Delta|^2} \qquad (10\text{-}18)$$

Radius of the circle located at r_{s2} is

$$R_{s2} = \frac{|S_{12}S_{21}|}{|S_{22}|^2 - |\Delta|^2} \qquad (10\text{-}19)$$

where

$$C_1 = S_{11} - \Delta S_{22}*$$ (10-20)

$$C_2 = S_{22} - \Delta S_{11}*$$ (10-21)

$$\Delta = S_{11}S_{22} - S_{12}S_{21}$$ (10-22)

* Denotes the complex conjugate.

GAIN

MAXIMUM STABLE GAIN (MSG)

$$\text{MSG} = \sqrt{\frac{\text{Forward Gain}}{\text{Reverse Gain}}} = \left| \frac{S_{21}}{S_{12}} \right|$$ (10-23)

MAXIMUM AVAILABLE GAIN (G_{max} OR MAG)

$$G_{max} = \text{MAG} = \text{MSG} (K - \sqrt{K^2 - 1})$$ (10-24)

$$= \left| \frac{S_{21}}{S_{12}} \right| (K - \sqrt{K^2 - 1})$$ (10-25)

UNILATERAL GAIN (G_u)

Reverse gain is assumed to equal 0.

$$G_u = \frac{1}{2K} \left[\frac{|(S_{21}/S_{12})^{-1}|^2}{|S_{21}/S_{12}| - R_e(S_{21}/S_{12})} \right]$$ (10-26)

$$= 0.5 \left[\frac{\text{MSG} - 2 \cdot \cos \theta + \text{MSG}^{-1}}{K - \cos \theta} \right]$$ (10-27)

$$\cong 0.5 \left(\frac{\text{MSG}}{K - \cos \theta} \right)$$ (10-28)

θ is the difference between the forward and reverse phase shift

R_e denotes "the real part of"

K is Rollett's Stability Factor

CONSTANT-GAIN CIRCLES

To find the center of the constant-gain circle,

$$r_{02} = \left[\frac{G}{1 + D_2 G} \right] C_2*$$ (10-29)

Radius of the circle:

$$R_{02} = \frac{(1 - 2K|S_{11}S_{21}|G + |S_{12}S_{21}|^2 G^2)^{\frac{1}{2}}}{1 + D_2 G}$$ (10-30)

K is the stability factor

$$G = G_p/|S_{21}|^2$$ (10-31)

G_p is the desired numeric gain

$$D_2 = |S_{22}|^2 - |\Delta|^2 \tag{10-32}$$

$$\Delta = S_{11}S_{22} - S_{12}S_{21} \tag{10-33}$$

Power Gain Less than G_{max}

Select a load on the constant gain circle to give the desired gain. Select a generator impedance to achieve this gain. The generator impedance matching the input load is

$$r_1 = \left[\frac{S_{11} - r_2\,\Delta}{1 - r_2 S_{22}} \right]^* \tag{10-34}$$

where

r_2 is the selected load reflection coefficient

$$\Delta = S_{11}S_{22} - S_{12}S_{21} \tag{10-35}$$

* Denotes the complex conjugate.

OPERATIONAL AMPLIFIERS

Common-Mode Rejection (CMR)

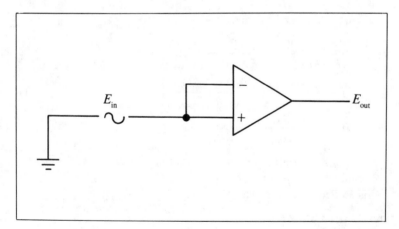

$$CMR = \frac{E_{out}}{E_{in}A_{ol}} \quad \text{(in } dB) \tag{10-36}$$

A_{ol} is the open-loop gain

$$CMR \text{ (dB)} = 20 \log_{10} CMR \tag{10-37}$$

Equivalent Signal Differential (ESD)

$$ESD = \frac{E_{out}}{A_{ol}} \tag{10-38}$$

A_{ol} is the open-loop gain

INPUT OFFSET VOLTAGE (V_{io})

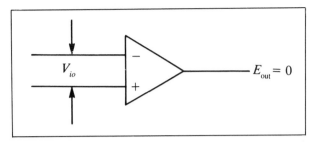

INPUT OFFSET CURRENT (I_{io})

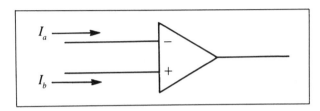

$$I_{io} = |I_a - I_b| \qquad (10\text{-}39)$$

INPUT BIAS CURRENT (I_{ib})

Refer to illustration above.

$$I_{ib} = \tfrac{1}{2}I_{io} \qquad (10\text{-}40)$$

Note: To minimize V_{io} and I_{io}, the positive input port series resistance R_2 should be made equal to the parallel resistance value of the negative input port series resistance and the feedback resistance.

AMPLIFIER CONFIGURATION

Approximate Relationships

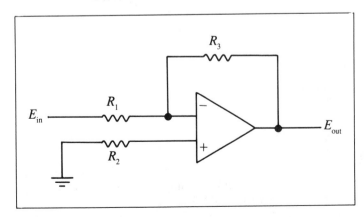

$$R_2 = \frac{R_1 R_3}{R_1 + R_3} \tag{10-41}$$

Voltage gain (A)

$$A = \frac{E_{out}}{E_{in}} \tag{10-42}$$

$$= \frac{R_3}{R_1} \tag{10-43}$$

Input impedance (Z_{in})

$$Z_{in} \approx R_1 + \frac{R_3}{A_{ol}} \tag{10-44}$$

$$\approx R_1$$

A_{ol} is the open-loop gain

Output impedance (Z_{out})

$$Z_{out} = \frac{Z_{out(ol)}}{1 + A_{ol}\left(\dfrac{R_1}{R_1 + R_3}\right)} \tag{10-45}$$

where

$Z_{out(ol)}$ is the open-loop output impedance

Unity-Gain, High-Input-Impedance Amplifier

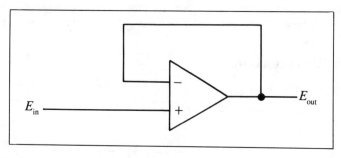

Gain (A)

$$A = \frac{E_{out}}{E_{in}} = 1 \tag{10-46}$$

Input Impedance (Z_{in})

$$Z_{in} = A_{ol}Z_{in(ol)} \tag{10-47}$$

A_{ol} and $Z_{in(ol)}$ are the open-loop gain and open-loop input impedance respectively.

Output Impedance (Z_{out})

$$Z_{out} = \frac{Z_{out(ol)}}{A_{ol}}$$ (10-48)

$Z_{out(ol)}$ is the open-loop output impedance

High-Input-Impedance Amplifier

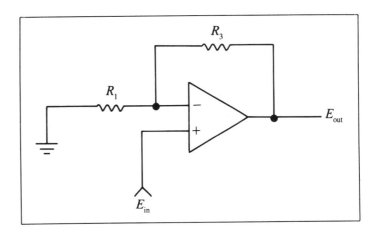

Input Impedance (Z_{in})

$$Z_{in} \cong AZ_{in(ol)}$$ (10-49)

$Z_{in(ol)}$ is the open-loop input impedance

Gain (A)

$$A = \frac{E_{out}}{E_{in}} = \frac{R_1 + R_3}{R_1}$$ (10-50)

Output Impedance (Z_{out})

$$Z_{out} = \frac{Z_{out(ol)}}{1 + [A_{ol}R_1/(R_1 + R_3)]}$$

Differential Amplifier

$$E_{out} = E_{in2}\left(\frac{R_2}{R_2 + R_4}\right)\left(\frac{R_1 + R_3}{R_1}\right) - E_{in1}\frac{R_3}{R_1}$$ (10-51)

When $R_1 = R_4$ and $R_2 = R_3$,

$$E_{out} = (E_{in2} - E_{in1})\frac{R_3}{R_1}$$ (10-52)

$$= E_{in2} - E_{in1})A$$ (10-53)

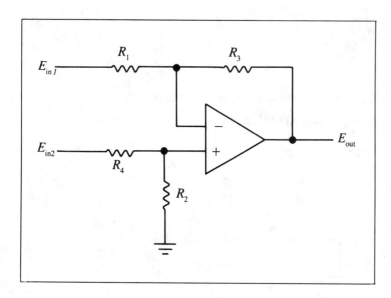

where

    A is gain and $A = R_3/R_1$

When $R_1 = R_2 = R_3 = R_4$,

    $E_{out} = E_{in2} - E_{in1}$ (10-54)

Summing Amplifier Inverter

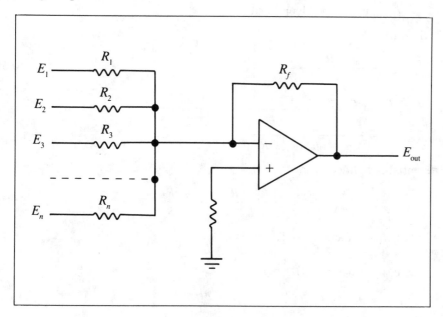

Output Voltage (E_out)

$$E_{out} = -R_f\left(\frac{E_1}{R_1} + \frac{E_2}{R_2} + \cdots \frac{E_n}{R_n}\right)$$ (10-55)

Channel Input Impedance (Z_n)

$$Z_n = R_n$$

Channel Gain (A_n)

$$A_n = \frac{R_f}{R_n}$$ (10-56)

Input Current (I_n)

$$I_n = E_n/R_n$$ (10-57)

Integrator

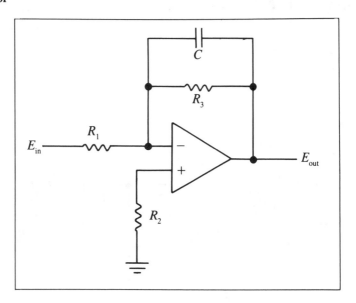

$$R_2 = \frac{R_1 R_3}{R_1 + R_3}$$ (10-58)

$$E_{out} = \frac{1}{R_1 C}\int E_{in}\, dt$$ (10-59)

Note: The output E_{out} is shifted in phase by 90°.

Differentiator

$$E_{out} = R_3 C\, \frac{\Delta E_{in}}{\Delta t}$$ (10-60)

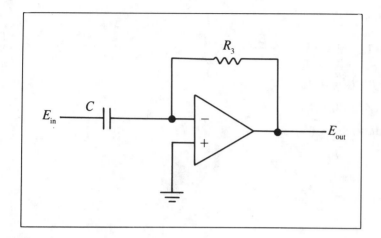

VIDEO AMPLIFIER WITH SERIES AND PARALLEL FEEDBACK

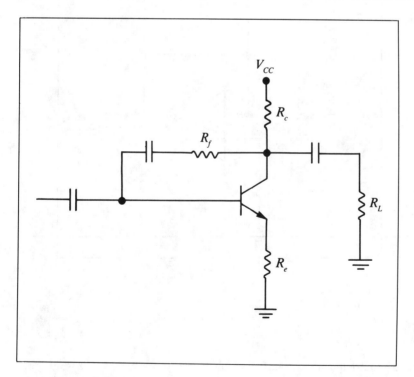

INPUT IMPEDANCE (Z_{in})

$$Z_{in} = R_e\left(\frac{R_f + R_c}{R_e + R_c}\right)$$

(10-61)

Output Impedance (Z_{out})

$$Z_{out} = R_e \left(\frac{R_f + R_s}{R_e + R_s} \right) \tag{10-62}$$

Voltage Gain (G_v)

$$G_v = - \left(\frac{R_c}{R_e} \right) \left(\frac{R_f - R_e}{R_c + R_f} \right) \tag{10-63}$$

If R_L is 50 Ω and is less than R_c, and if $R_e R_f = 50^2$, then

$$Z_{in} \cong Z_{out} \cong 50; \text{ VSWR} \approx 2:1$$

FIELD-EFFECT TRANSISTORS

Amplifiers

Common-Source Configuration

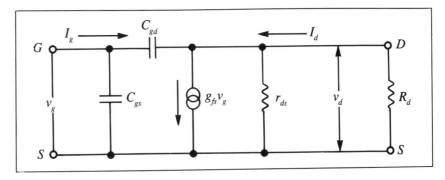

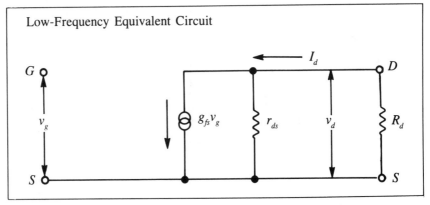

Low-Frequency Equivalent Circuit

g,G is gate
d,D is drain
s,S is source

g_{fs} is transconductance
C is capacitance
I is current
v is voltage
r and R are resistances

Drain Current

$$I_d = g_{fs}v_g r_{ds} \frac{1}{r_{ds} + R_d} \tag{10-64}$$

Output Voltage

$$v_o = g_{fs}v_g \frac{r_{ds}R_d}{r_{ds} + R_d} \tag{10-65}$$

Voltage Gain

$$A_v = g_{fs} \frac{r_{ds}R_d}{r_{ds} + R_d} \tag{10-66}$$

If $r_{ds} > R_d$,

$$A_v = g_{fs}R_d \tag{10-67}$$

Output Resistance

$$R_o = \frac{r_{ds}R_d}{r_{ds} + R_d} \tag{10-68}$$
$$\approx R_d \text{ if } r_{ds} > R_d$$

Input resistance is usually determined by the bias network at the gate.

Common-Source Amplifiers With Source Feedback at Low Frequencies

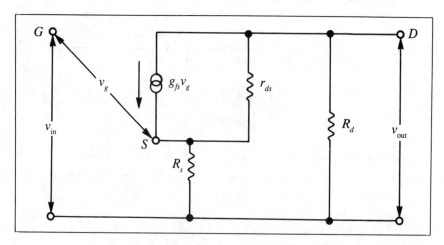

Gate Input Voltage

$$v_g = v_{in} - I_d R_s \qquad (10\text{-}69)$$

Drain Current

$$I_d = g_{fs} v_{in} r_{ds} \frac{1}{r_{ds} + R_d + R_s + g_{fs} R_s r_{ds}} \qquad (10\text{-}70)$$

Output Voltage

$$v_{out} = I_d R_d \qquad (10\text{-}71)$$

Voltage Gain

$$A_v = g_{fs} r_{ds} R_d \frac{1}{r_{ds} + R_d + R_s + g_{fs} R_s r_{ds}} \qquad (10\text{-}72)$$

Source Follower (Common Drain)

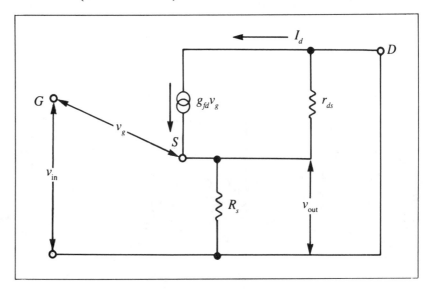

Gate Voltage

$$v_g = v_{in} - R_s I_d \qquad (10\text{-}73)$$

Drain Current

$$I_d = g_{fd} v_g + \frac{V_{out}}{r_{ds}} \qquad (10\text{-}74)$$

Output Voltage

$$V_{out} = \left(g_{fd} v_g + \frac{v_{out}}{r_{ds}} \right) R_s \qquad (10\text{-}75)$$

$$= \frac{v_{in} g_{fd} r_{ds} R_s}{r_{ds} + g_{fd} r_{ds} R_s + R_s} \tag{10-76}$$

Voltage Gain

$$A_v = g_{fd} r_{ds} R_s \frac{1}{r_{ds} + g_{fd} r_{ds} R_s + R_s} \tag{10-77}$$

$$\approx g_{fd} R_s (1 + g_{fd} R_s) \quad \text{(for } r_{ds} > R_s) \tag{10-78}$$

Approximations

Each of the three common configurations are shown schematically, followed by their corresponding gain and impedance formula approximations.

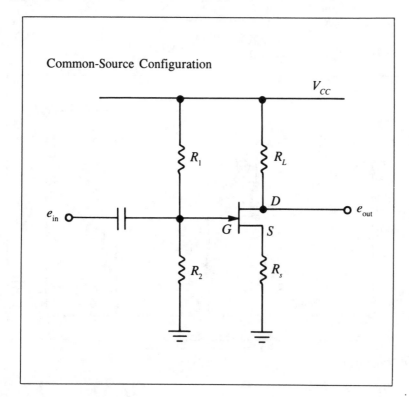

Common-Source Configuration

Common-Source Voltage Gain

$$G_v \approx \frac{-R_L}{\dfrac{1}{g_m} + R_S} \tag{10-79}$$

Output Impedance

$$Z_{out} \approx R_L \tag{10-80}$$

Input Impedance

$$Z_{in} \approx \frac{R_1 R_2}{R_1 + R_2} \tag{10-81}$$

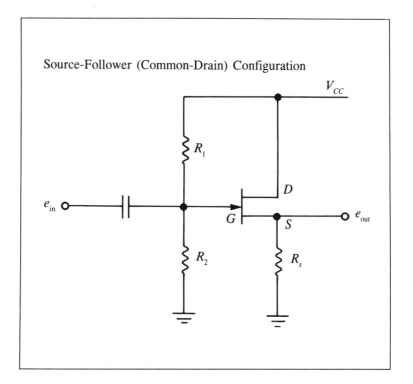

Source-Follower (Common-Drain) Configuration

Source-follower Voltage Gain

$$G_v \approx \frac{R_s}{\dfrac{1}{g_m} + R_s} \tag{10-82}$$

Output Impedance

$$Z_{out} \approx \frac{R_s \dfrac{1}{g_m}}{R_s + \dfrac{1}{g_m}} \tag{10-83}$$

Input Impedance

$$Z_{in} \approx \frac{R_1 R_2}{R_1 + R_2} \tag{10-84}$$

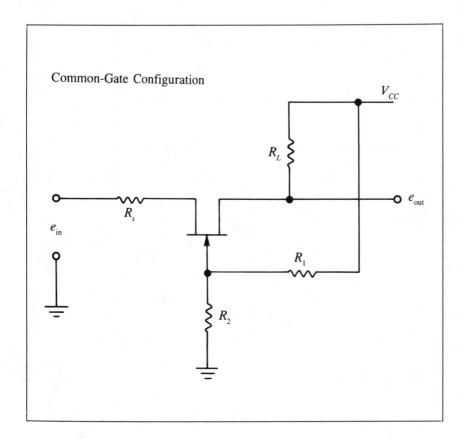

Common-Gate Configuration

Common-Gate Voltage Gain

$$G_v \approx \frac{R_L}{\dfrac{1}{g_m} + R_s}$$

(10-85)

Output Impedance

$$Z_{\text{out}} \approx R_L$$

(10-86)

Input Impedance

$$Z_{\text{in}} \approx R_s + \frac{1}{g_m}$$

(10-87)

CASCADE SYNCHRONOUS SINGLE-TUNED AMPLIFIER BANDWIDTH

Approximate Bandwidth (Overall)

$$B = \text{One-stage Bandwidth}/(1.2\sqrt{n})$$

n is the number of stages

Exact Overall Bandwidth

$$B = \text{One-stage Bandwidth} \times (2^{1/n} - 1)^{\frac{1}{2}}$$

EXAMPLE:

For a bandwidth of 1 MHz in 6 stages, the single-stage bandwidth must be 1/.35 or 2.8 MHz.

11

noise

THERMAL NOISE

MEAN SQUARE NOISE VOLTAGE

$$\overline{e^2} = 4\,k\,T\,R\,B \tag{11-1}$$

k is Boltzman's constant ($1.38 \cdot 10^{-23}$ joules per degree Kelvin)
T is the temperature in degrees Kelvin
R is resistance (in *ohms*)
B is the effective noise bandwidth (in *Hz*)

MAXIMUM AVAILABLE NOISE POWER FOR CONJUGATE MATCHING

$$P_n = k\,T\,B \tag{11-2}$$

KTB

EXAMPLE:

If $B = 20$ kHz, then $kTB = -134 + 3 = -131$ dBm

EFFECTIVE NOISE TEMPERATURE (T_e)

$$T_e = (F - 1)T_o \tag{11-3}$$

where
F is noise factor
T_o is the temperature of the device

145

Table 11-1. Values for kTB

$k = 1.38 \cdot 10^{-23}$ watts per degree Kelvin $T = 290$ degrees Kelvin B is bandwidth (in Hz)	
B	**kTB (dBm)**
1 Hz	-174
10 Hz	-164
100 Hz	-154
1000 Hz	-144
10 kHz	-134
100 kHz	-124
1MHz	-114

Note: F and T_o must be referenced to the same temperature, usually taken as 290°K (room temperature).

CASCADE NOISE TEMPERATURE

$$T_{e_{\text{total}}} = T_1 + \frac{T_2}{G_1} + \frac{T_3}{G_1G_2} + \cdots \frac{T_n}{G_1G_2 \cdots G_{n-1}} \qquad (11\text{-}4)$$

where
  n refers to the stage in the network

TRANSMISSION-LINE EFFECTIVE NOISE TEMPERATURE

$$T_e = T_a(L - 1) \qquad (11\text{-}5)$$

where
  L is the loss of the line
  T_a is the ambient temperature (in *degrees Kelvin*)

ATMOSPHERIC LOSS AND NOISE TEMPERATURE

$$T_e = T_a(L - 1) \qquad (11\text{-}6)$$

where
  L is the atmospheric loss (ratio)
  T_a is the ambient temperature (in *degrees Kelvin*)

NOISE TEMPERATURE OF LOSSY COMPONENTS

$$T_e = T_a(L - 1) \qquad (11\text{-}7)$$

where
  L is the loss of the component equal to the reciprocal of the fractional gain
  T_a is the ambient temperature (in *degrees Kelvin*)

EXAMPLE:

Gain = 0.25, $L = 4$, $T_a = 290°K$, and $T_e = 870°K$.

CASCADE NOISE FACTOR

TWO NETWORKS

$$F_{total} = F_1 + \frac{F_2 - 1}{G_1} \tag{11-8}$$

where
F_1 is the first-stage noise factor
F_2 is the second-stage noise factor
G_1 is the first-stage gain

Note: This equation may be repeatedly applied to a cascade of networks of greater than two by starting at the output end as the second stage and the preceding stage as stage 1. The resulting value of F is then used as the second stage F and the preceding stage is stage 1, etc.

NOISE FACTOR OF LOSSY COMPONENTS

$$F = L \tag{11-9}$$

where
L is the reciprocal of the fractional gain G

EXAMPLE:

If $G = 0.25$, then L equals a ratio of 4 and $F = 4$.

NOISE FIGURE OF LOSSY COMPONENTS

$$NF = L(dB) \tag{11-10}$$

EXAMPLE:

If $G = 0.25$, then L equals a ratio of 4, which is 6 dB. Therefore, $NF = 6$ dB.

NOISE FACTOR OF NETWORKS (GENERAL CASE)

$$F_{total} = F_1 + \frac{F_2 - 1}{G_1} + \frac{F_3 - 1}{G_1 G_2} + \frac{F_4 - 1}{G_1 G_2 G_3} +$$
$$\cdots \frac{F_n - 1}{G_1 G_2 G_3 G_4 \cdots G_{n-1}} \tag{11-11}$$

The subscripts denote the stage involved.

NOISE FACTOR FROM NOISE TEMPERATURE

$$F = 1 + T_e/T_o \tag{11-12}$$

where
  T_e is the effective noise temperature
  T_o is 290°K

NOISE FIGURE (*NF*)

$$NF = 10 \log_{10} F \quad \text{(in } dB\text{)} \tag{11-13}$$

NOISE FACTOR (*F*)

$$F = \frac{S_{in}/N_{in}}{S_{out}/N_{out}} \tag{11-14}$$

where
  S_{in}/N_{in} is the signal-to-noise power ratio at the input
  S_{out}/N_{out} is the output signal-to-noise power ratio
  $N_{in} = kTB$

Or,

$$F = \frac{S_{in}N_{out}}{S_{out}N_{in}} = \frac{N_{out}}{GN_{in}} \tag{11-15}$$

  G is gain

COMPUTING NOISE FIGURE TO AND FROM TANGENTIAL SIGNAL SENSITIVITY (TSS)

Noise Figure from Tangential Signal Sensitivity

$$NF = \text{Tangential Signal Sensitivity} - kTB - 8 \quad \text{(in } dB\text{)} \tag{11-16}$$

Tangential Signal Sensitivity from Noise Figure

$$\text{Tangential Signal Sensitivity} = NF + kTB + 8 \quad \text{(in } dBm\text{)} \tag{11-17}$$

where
  Tangential Signal Sensitivity is the tangential sensitivity
  NF is noise figure
  kTB is noise in a bandwidth B
    $= -114$ dBm per MHz
    $= -144$ dBm per kHz
    $= -174$ dBm per Hz

SIGNAL-TO-NOISE RATIO

$$S/N = \frac{\text{Signal Power}}{\text{Noise Power}} \tag{11-18}$$

SIGNAL-PLUS-NOISE-TO-NOISE RATIO

$$\frac{S + N}{N} = \frac{\text{Signal Power} + \text{Noise Power}}{\text{Noise Power}} \tag{11-19}$$

In dB notation,

$$10 \log_{10} (S + N) - 10 \log_{10} (N) \tag{11-20}$$

S/N TO $(S + N)/N$ CONVERSION

$$(S/N) + 1 = (S + N)/N \tag{11-21}$$

When S/N is 20 dB or more,

$$S/N = (S + N)/N \tag{11-22}$$

SINAD

Sinad stands for "signal plus noise plus distortion to noise plus distortion" ratio. As the name implies, the total receiver output (signal, noise, and distortion) is compared to the same input without the signal portion. In other words, the signal modulation is nulled out in the denominator of this ratio, leaving the noise and distortion.

$$\text{sinad} = \frac{S + N + D}{N + D} \tag{11-23}$$

where
   S is signal power
   N is noise power
   D is distortion power

Measurement:

- Measure the power of $S + N + D$
- Measure the power of $N + D$ by nulling out S
- Compute the power ratio for sinad

RECEIVER OUTPUT SIGNAL-TO-NOISE RATIO

S_{out}/N_{out} is defined as the receiver's output signal relative to its normal, no-signal noise level. This ratio is governed by the required threshold for signal recognition.

$$S_{out}/N_{out} = \frac{S_{in}}{N_{in}F} = S_{in}/(kTBF) \tag{11-24}$$

INPUT SIGNAL STRENGTH FOR A GIVEN S_{out}/N_{out}

$$S_{in} = (S_{out}/N_{out})kTBF \tag{11-25}$$

Then the minimum input signal is

$$S_{in_{min}} = (S_{out}/N_{out})_{min}kTBF \tag{11-26}$$

BIT-ERROR PROBABILITY RELATED TO S/N

On/Off Keying

$$P_e = \tfrac{1}{2}erfc\tfrac{1}{2}(S/N)^{\frac{1}{2}} \tag{11-27}$$

Amplitude Shift Keying (Incoherent)

$$P_e = \tfrac{1}{2}e^{-S/(2N)}\left[1 + \frac{1}{(2\pi S/N)^{\frac{1}{2}}}\right] \tag{11-28}$$

Differential Phase-Shift Keying

$$P_e = \tfrac{1}{2}e^{-S/N} \tag{11-29}$$

Frequency-Shift Keying

$$P_e = \tfrac{1}{2}e^{-S/(2N)} \tag{11-30}$$

Polar Pulse-Code Modulation and Phase-Shift Keying

$$P_e = \tfrac{1}{2}erfc(S/N)^{\frac{1}{2}} \tag{11-31}$$

Unipolar Pulse-Code Modulation and Coherent Amplitude and Frequency Shift Keying

$$P_e = \tfrac{1}{2}erfc(S/(4N))^{\frac{1}{2}} \tag{11-32}$$

where

  P_e is the bit-error probability
  $erfc$ is the complementary error function
  S is signal power
  N is noise power

ENERGY-PER-BIT-TO-NOISE DENSITY (e_b/N_o) _____

$$e_b/N_o = 20\log_{10}(e_s/e_n) + 10\log_{10}(ENB/b) \qquad \text{(in } dB\text{)} \tag{11-33}$$

where

  e_s is the signal magnitude measured with a true rms meter
  e_n is the noise measured with a true rms meter
  ENB is the effective noise bandwidth
  b is the bit rate

The energy per bit is

$$e_b = (e_s{}^2/z)(1/b) \tag{11-34}$$

The noise density is

$$N_o = (e_n^2/z)(1/ENB) \qquad (11\text{-}35)$$

z is impedance

EFFECTIVE NOISE BANDWIDTH (*ENB*)

The *ENB* of a network or filter is a rectangle of equivalent area and height of the power versus frequency response of that network or filter.

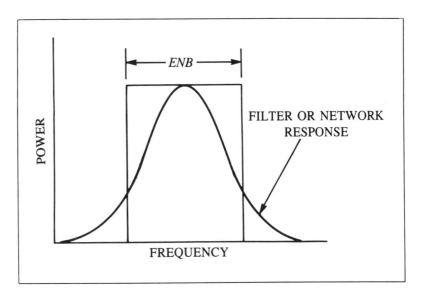

POWER LINE FIELD STRENGTH

$$B = 0.2I/D \qquad (\text{in } gausses) \qquad (11\text{-}36)$$

D is distance (in *cm*)
I is current (in *amperes*)

12

---

antennas

EFFECTIVE AREA OF AN ISOTROPIC ANTENNA ———————————

$$A_e = \lambda^2/(4\pi) \tag{12-1}$$

where

λ is the operating wavelength

Note: The gain of an isotropic antenna is considered to be unity. Other antennas are referred to this antenna as a reference and are noted in dB_i (dB as referred to the isotropic case).

ANTENNA EFFECTIVE AREA ———————————

The effective area of any antenna is

$$A_e = \frac{G\lambda^2}{4\pi} \tag{12-2}$$

where

G is the gain of the antenna as referred to the isotropic case

GAIN MEASUREMENT: TWO-ANTENNA SYSTEM ———————————

Two identical antennas are spaced apart by a distance D where

$$D > 2A^2/\lambda \tag{12-3}$$

A is the maximum aperture of the antenna.

One antenna is the transmitter and the other is the receiving antenna. The antenna gain is

$$G_{dB} = 10 \log_{10} (4\pi D/\lambda)(P_r/P_t)^{\frac{1}{2}} \tag{12-4}$$

P_r is watts received
P_t is watts transmitted
λ is wavelength

VERTICAL ANTENNAS

ANTENNA ABOVE A GROUND PLANE

The electric field is

$$E(\phi) = \frac{60\,I}{r} \frac{\cos\left(\dfrac{2\pi l}{\lambda}\sin\phi\right) - \cos\left(\dfrac{2\pi l}{\lambda}\right)}{\cos\phi} \tag{12-5}$$

where
  ϕ is the vertical angle relative to the ground plane
  l is current
  r is range
  λ is wavelength
  l is length

SHORT VERTICAL ANTENNA (h < λ/4)

Impedance

$$Z_o = 60\left(\ln\frac{2h}{a} - 1\right) \tag{12-6}$$

h is the antenna height above a ground plane in degrees as referred to a wavelength at the operating frequency
a is the diameter of the vertical antenna in degrees as referred to a wavelength

Top Loading by a Disk of Radius r

$$r = (7.5\lambda \tan H_b)Z_o \tag{12-7}$$

$$H_b = h_{\lambda/4} - h \tag{12-8}$$

$h_{\lambda/4}$ is the height of a $\lambda/4$ antenna
$\lambda = 360$ degrees (one wavelength)

Variables r, H_b, and $h_{\lambda/4}$ are in degrees in terms of λ.

Radiation Resistance

$$R_r = 40(1 - 0.085 \sin^{\frac{5}{2}} h) \sin^2 h \quad (in \; ohms) \tag{12-9}$$

QUARTER WAVE VERTICAL ANTENNA

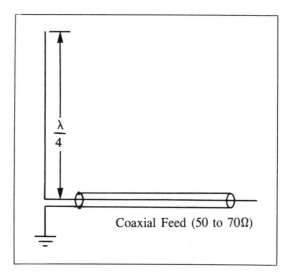

The polarization of this antenna is vertical. The gain can be raised by 3 dB by increasing its length to $\frac{5}{8}\lambda$.

GROUND-PLANE ANTENNA

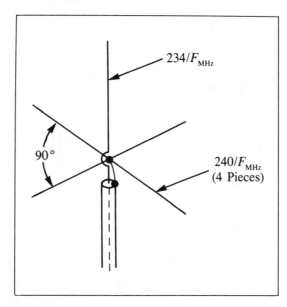

DISCONE ANTENNA

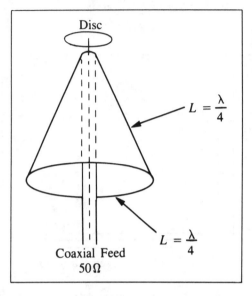

SWR is 1.5 to 1.0 over several frequency octaves. Polarization is vertical. The diameter (d) of the disc is $0.67L < d < 0.70L$.

COAXIAL DIPOLE

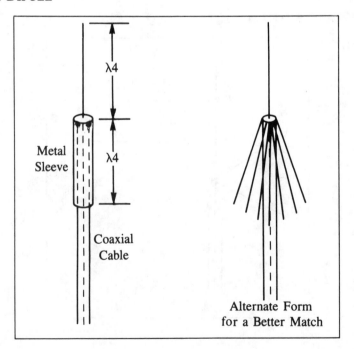

J ANTENNA

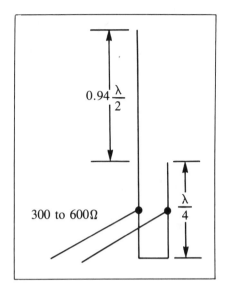

The base of the antenna may be grounded. The feed lines should be balanced, or they may be unbalanced lines with a balun. Straight coaxial feed can be used at the antenna base as in a conventional vertical antenna.

HORIZONTAL ANTENNAS

DIPOLE ANTENNAS

Length of a Half Dipole

$$L = 468/f_{\text{MHz}} \quad \text{(in } feet\text{)}\tag{12-10}$$

Impedance ($l = \frac{1}{2}\lambda$)

$$Z \cong 72 \text{ ohms}$$

Radiation Resistance—General Case

$$R_r = 789(l/\lambda)^2 \quad \text{(in } ohms\text{)}\tag{12-11}$$

l is length
λ is wavelength (identical units)

Electric Field (in the Dipole Plane)

$$E(\theta) = \frac{60\pi ll}{r\lambda}\cos\theta \quad \text{(in } volts\ per\ meter\text{)}\tag{12-12}$$

l is length
I is current

r is distance

λ is wavelength

θ is the angle from the dipole perpendicular

Note: This equation describes the field of a horizontal dipole in the horizontal plane or a vertical dipole in the vertical plane.

FOLDED DIPOLES

The impedance of a folded dipole increases as the folds or parallel elements are squared.

Design

This antenna form can be made from a parallel open transmission line such as TV twinlead, as shown.

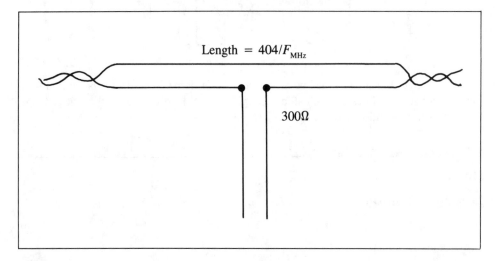

Length = $404/F_{MHz}$

300Ω

HORIZONTAL ANTENNA *n* WAVELENGTHS LONG

Resonant Wire in Free Space *n*λ Long and End-Fed

With n Even (Electric Field)

$$E(\theta) = (60I/r)\left[\frac{\sin\left(\dfrac{\pi l}{\lambda}\cos\theta\right)}{\sin\theta}\right] \quad (in\ volts\ per\ meter) \qquad (12\text{-}13)$$

With n Odd

$$E(\theta) = (60I/r)\left[\frac{\cos\left(\dfrac{\pi l}{\lambda}\cos\theta\right)}{\sin\theta}\right] \quad (in\ volts\ per\ meter) \qquad (12\text{-}14)$$

where

    θ is the angle relative to the axis

    I is current

    l is length

    λ is wavelength

    r is distance

CENTER-FED WIRE

For any length antenna, horizontal or vertical, in free space (electric field):

$$E(\theta) = \frac{60I}{r} \; \frac{\cos\left(\dfrac{\pi l}{\lambda}\cos\theta\right) - \cos\left(\dfrac{\pi l}{\lambda}\right)}{\sin\theta} \qquad (12\text{-}15)$$

where

    θ is the angle relative to ths axis

    I is current

    l is length

    λ is wavelength

    r is distance

CENTER-FED HARMONIC DOUBLET

$$\text{Length} = \frac{(K - 0.05)492}{f_{\text{MHz}}} \qquad (\text{in } feet) \qquad (12\text{-}16)$$

K is the number of $\frac{1}{2}$ waves

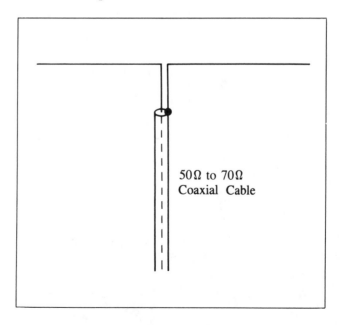

50 Ω to 70 Ω
Coaxial Cable

Note: This is a balanced antenna that should be fed by a balanced line or balun.

BEVERAGE ANTENNA

The beverage antenna is a terminated nonresonant horizontal wire above ground. For the electric field,

$$E(\theta) = \frac{60I}{r} \frac{\sin \theta \sin \left[\dfrac{\pi l}{\lambda}(1 - \cos \theta) \right]}{1 - \cos \theta} \tag{12-17}$$

where
θ is the angle relative to the axis
I is current
r is range
λ is wavelength
l is length of the wire

Note: The antenna is terminated at the wire end away from the transmitter or receiver and has vertical polarization.

HALF-WAVE ANTENNA

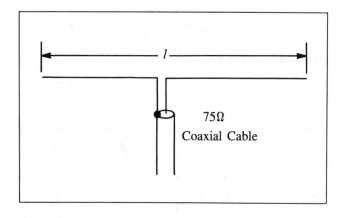

$$l = \frac{468}{f_{\text{MHz}}} \quad (\text{in } feet)$$

FOUR-LEAF CLOVER

This antenna consists of two dipoles in space quadrature resulting in an antenna pattern that forms a four-leaf clover in geometry.

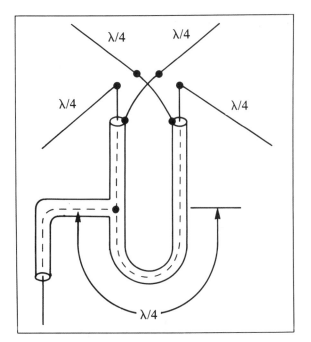

PARABOLIC REFLECTORS

DEFINITIONS

The parabola is a special case of a conic described by

$$x^2 + y^2 = e^2(d + x)^2 \qquad (12\text{-}18)$$

where

d is the distance from the focus f to the directrix (a vertical line)

e is eccentricity and is the ratio of the distance d_1 of a point p to the focus f to the distance d_2 of the point to the directrix by shortest route (horizontal)

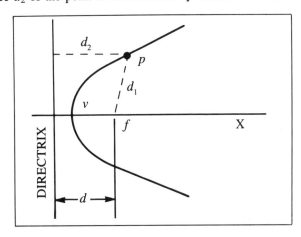

To go from conic to parabolic, let $d = \frac{1}{2}a$ (where a is a distance along x) and $e = 1$. Then

$$x^2 + y^2 = (\tfrac{1}{2}a + x)^2 \tag{12-19}$$

or

$$y^2 = ax \tag{12-20}$$

Additionally the latus rectum (LR) is equal to a and the distance of the focus f to the vertex v is $\frac{1}{4}a$:

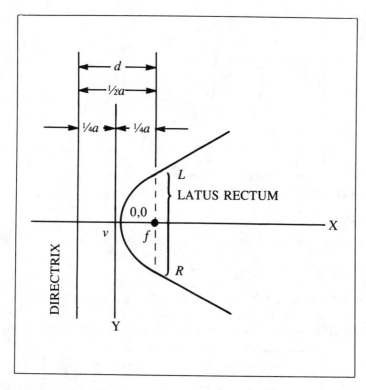

GAIN (REFERRED TO A DIPOLE)

$$G \approx 6(D\lambda)^2 \tag{12-21}$$

D is the diameter of the parabola
λ is the wavelength (in compatable units)

BEAM WIDTH (B)

$$B = 58k \frac{\lambda}{D} \quad \text{(in } \textit{degrees}\text{)} \tag{12-22}$$

k is the illumination factor, which is 1 for the ideal case and $1\frac{1}{4}$ for the usual case

FOCAL LENGTH (f)

$$f = D^2/16C \qquad\qquad (12\text{-}23)$$

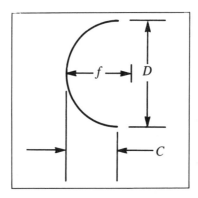

PARABOLIC REFLECTOR SPINCASTING

Spincasting has an accuracy suitable for optical as well as radio-frequency purposes. A suitable material (such as epoxy and metal particle mixes) when placed in a container and spun will assume a parabolic shape. When this is allowed to harden, a parabolic reflector results. The relations of the parameters involved is:

$$f = (450g/\pi^2 n^2) \qquad\qquad (12\text{-}24)$$

where

f is the focal length of the parabola desired (in *feet* or *cm*)
g is the gravity at the location (in *feet/sec²* or *cm/sec²*)
π is radians
n is the rotation per minute

MICROWAVE ANTENNAS

HORN ANTENNA DESIGN

Solve for l_e (the E-plane) or l_h (the H-plane apertures) or both using the relationship for the desired beam width in the following sections.

Solve for L (the horn length) from

$$L = \frac{\left[\left(\dfrac{1}{2\lambda}\right)^2 - \delta^2\right]\lambda}{2\delta} \qquad\qquad (12\text{-}25)$$

where

δ is the path length difference in wavelengths
$\delta = 0.25$ for optimum E-plane horns
$\delta = 0.40$ for optimum H-plane horns
$\delta = 0.32$ for conical horns

Solve for the horn flare angle θ from

$$\theta = 2 \cos^{-1} \left(\frac{L/\lambda}{L/\lambda + \delta} \right) \qquad (12\text{-}26)$$

GAIN

$$G = 7.5 \frac{l_e l_h}{\lambda^2} \qquad (12\text{-}27)$$

SECTORAL HORN ANTENNAS

For optimum configurations, beam width, length of side lobes, and phase error are compromised.

Configuration of E-plane Horn

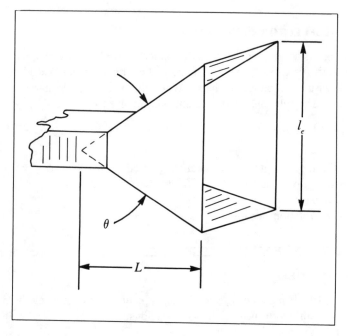

Beam Width (Half Power)

$$B_{e\frac{1}{2}} = 56 \frac{\lambda}{l_e} \qquad (12\text{-}28)$$

Null Beam Width (Between Nulls)

$$B_{en} = 115 \frac{\lambda}{l_e} \qquad (12\text{-}29)$$

For definitions of L and θ, see equations (12-25) and (12-26), respectively.

CONICAL HORNS

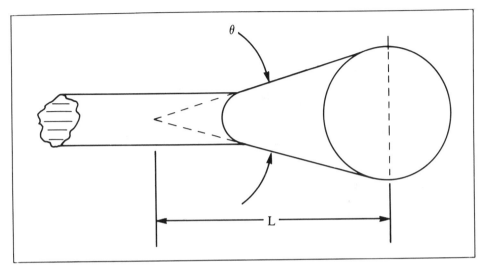

PYRAMIDAL HORN

A pyramidal is a combination of E- and H-plane horns. This is the most common of horn designs.

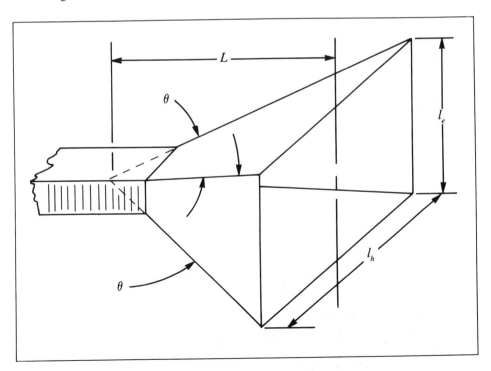

H-Plane Sectoral Horn

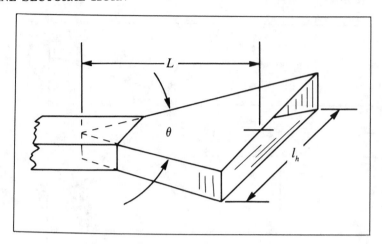

Beam Width (Half Power)

$$B_{h\frac{1}{2}} = 67\,\frac{\lambda}{l_h}$$

(12-30)

Null Beam Width (Between Nulls)

$$B_{hn} = 172\,\frac{\lambda}{l_h}$$

(12-31)

For definitions of L and θ, see equations (12-25) and (12-26), respectively.

MISCELLANEOUS ANTENNAS

Helical Antenna

The electric field is defined by

$$E(\theta) = \left(\sin\frac{90}{n}\right)(\cos\theta)\left(\frac{\sin\dfrac{n\alpha}{2}}{\sin\dfrac{\alpha}{2}}\right)$$

(12-32)

where

$$\alpha = 360\left[s(1 - \cos\theta)(1/\lambda) + \frac{1}{2n}\right]$$

(12-33)

θ is the angle relative to the helix axis
n is the number of turns in the helix
λ is wavelength
s is helix turn spacing

LOOP ANTENNAS

For such an antenna where the loop perimeter is small (less than the wavelength), the electric field is

$$E(\theta) = \frac{120\pi^2 AI}{r\lambda^2} \sin \theta \tag{12-34}$$

where

    θ is the angle relative to the loop face
    A is the loop area
    r is range
    I is current
    λ is wavelength

SLOT ANTENNA

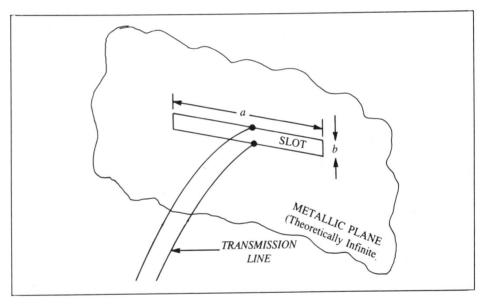

Slot Impedance

$$Z_s = (60\pi)^2/Z_{\text{dipole}} \tag{12-35}$$

$$a > 20b \tag{12-36}$$

Z_{dipole} is the impedance of a dipole having the dimensions of the slot. For a half-wave dipole, Z_{dipole} is 73 ohms.

For the condition stated in equation (12-36),

$$Z_s = 487 \text{ ohms}$$

and

$$a = 468/f(\text{MHz}) \tag{12-37}$$

POLARIZATION LOSS_____

Table 12-1.   Loss Between a Transmitting and a Receiving Antenna due to Polarization Effects

Antenna 1 Polarization	Antenna 2 Polarization (values in dB)			
	Circular			
	Right	Left	Vertical	Horizontal
Circular				
Right	0	∞	3	3
Left	∞	0	3	3
Vertical	3	3	0	∞
Horizontal	3	3	∞	0

13

radio frequencies

RADAR

RANGE TO TARGET

$$R = \tfrac{1}{2} \Delta tc \qquad\qquad (13\text{-}1)$$

Where

    c is the velocity of propagation $(3 \cdot 10^8$ meters per second)

    Δt is the round-trip time of the radar signal reflected from a target

Table 13-1. R Values per μs

0.081 nautical mile
0.093 statute miles
164 yards
492 feet
149.96 meters
0.1496 kilometers

Table 13-2. Range Equivalents

Units	English	Metric
Statute Mile	5280 ft	1609.3440 m
Nautical Mile	6076 ft	1851.9648 m
Radar Mile	2000 yds	1828.8000 m

UNAMBIGUOUS RANGE

$$R_u = 0.5 \frac{c}{prf}$$

(13-2)

where

R_u is the maximum unambiguous range

c is the velocity of propagation ($3 \cdot 10^8$ meters per second)

prf is the pulse repetition frequency (Hz)

THE EFFECTIVE RADIATED POWER (*ERP*)

$$ERP = P_t G_t$$

(13-3)

where

P_t is the transmitted power

G_t is the gain of the transmitting antenna

POWER DENSITY AT A DISTANCE *R*

$$P_d = P_t G_t \frac{1}{4\pi R^2}$$

(13-4)

POWER DENSITY AT THE RECEIVING ANTENNA

$$P_d{}' = P_t G_t \sigma [1/(4\pi R^2)]^2$$

(13-5)

σ is the radar cross section of the target.

POWER RECEIVED BY A RADAR ANTENNA OF AREA A_r

$$P_r = P_d{}' A_r = P_t G_t A_r \sigma [1/(4\pi R^2)]^2$$

(13-6)

ANTENNA GAIN-TO-AREA RELATIONSHIPS

$$G = \frac{4\pi A}{\lambda^2}$$

(13-7)

λ is the wavelength

or

$$A = \frac{G\lambda^2}{4\pi}$$

(13-8)

COMMON TRANSMIT/RECEIVE ANTENNA

$$G_t = G_r$$

(13-9)

$$A_t = A_r$$

(13-10)

Then

$$P_r = (P_t A^2 \sigma)/(4\pi\lambda^2 R^4)$$

(13-11)

and

$$P_r = (P_t G^2 \lambda^2 \sigma)/((4\pi)^3 R^4)$$ (13-12)

RADAR RANGE MAXIMUM

$$R_{\max} = \left[\frac{P_t A^2 \sigma}{(4\pi)P_{r_{\min}} \lambda^2} \right]^{\frac{1}{4}}$$ (13-13)

or

$$R_{\max} = \left[\frac{P_t G^2 \lambda^2 \sigma}{(4\pi)^3 P_{r_{\min}}} \right]^{\frac{1}{4}}$$ (13-14)

where

$P_{r_{\min}}$ is the minimum detectable signal power:

$$P_{r_{\min}} = S_{i_{\min}} \quad (\text{in } watts)$$

If $S_{i_{\min}} = $ Noise (rms),

$$R_{\max} = \left[\frac{P_t G_t G_r \lambda^2 \sigma}{(4\pi)^3 kTBF} \right]^{\frac{1}{4}}$$ (13-15)

k is Boltzman's constant ($1.38 \ 10^{-23}$ joules per degree Kelvin)
T is the temperature °Kelvin
B is the effective IF noise bandwidth
P_t is the transmitted power
G_t is the gain of the transmitting antenna
λ is the wavelength
σ is the radar cross section of the target
F is noise factor

RADAR CROSS SECTION (σ)

Sphere

$$\sigma = \pi r^2$$ (13-16)

$r = $ radius

Cone Axial Exposure

$$\sigma = 0.01989 \ \lambda^2 \ (\tan \theta)^4$$ (13-17)

$\theta = \frac{1}{2}$ of the conical angle

Paraboloid—Axial Exposure

$$\sigma = 4\pi C^2$$ (13-18)

$2C = $ radius of apex curvature

Prolate Spheroid—Axial Exposure

$$\sigma = \pi A_n^{\ 4}/A_j^{\ 2}$$ (13-19)

A_n = axis semiminor
A_j = axis semimajor

MINIMUM DISCERNABLE SIGNAL

Radar Receiver Noise (N_o)

$$N_o = kTBGF \tag{13-20}$$

T is the temperature in degrees Kelvin (usually 290°K)
K is 1.38 10^{-23} joules per degree Kelvin
B is the effective noise bandwidth of the receiver (ENB)
G is receiver gain
F is the receiver noise factor

Note: ENB is a rectangular passband whose power area is equal to that of the receiver filter.

Receiver Noise Figure (NF) and Noise Factor (F)

$$NF = 10 \log_{10} F \tag{13-21}$$

$$F = \frac{N_o}{kTBG} = \frac{S_{in}/N_{in}}{S_{out}/N_{out}} \tag{13-22}$$

S_{in}/N_{in} is the input signal-to-noise ratio
S_{out}/N_{out} is the output signal-to-noise ratio

Radar Post-Detection Pulse Integration

These equations define the integration improvement factor.

Ideal Integrator

$$I_{if} = nE_i(n) \tag{13-23}$$

where
   n is the number of pulses integrated
   $E_i(n) = (S/N)_1/[n(S/N)_n]$
   $(S/N)_1$ is the single-pulse signal-to-noise ratio for a given P_d (probability of detection)
   $(S/N)_n$ is the signal-to-noise ratio resulting from the integration of n pulses for the P_d above

Nonideal Case

$$I_{if} = \ln E_i(n) \tag{13-24}$$

where
   l is the nonideal loss factor
      $l = 0.10$ to 0.15
         $0.50 < P_d < 0.99$
         $10^{-12} < T_{fa} < 10^{-4}$

For the visual case, $l = 0.025$

where

> P_d is the probability of detection
> T_{fa} is the average false-alarm time

Time between False Alarms

$$T_{fa} = \frac{1}{B_{if}} \exp\left(\frac{E_t^2}{2n_o}\right) \quad \text{(in } seconds\text{)}$$

(13-25)

where

> B_{if} is the IF bandwidth (in Hz)
> E_t is the threshold level (in $volts$)
> n_o is the rms noise voltage

Probability of False Alarm

$$P_{fa} = \exp\left(-\frac{E_t^2}{2n_o}\right)$$

(13-26)

FM CW RADAR

LINEAR MODULATION

Beat Frequency (f_b)

$$f_b = 4Rf_m \frac{\Delta f}{c}$$

(13-27)

where

> R is range
> Δf is the FM deviation, peak to peak
> f_m is the FM rate
> $c = 3 \cdot 10^8$ meters per second

Range (R)

$$R = \tfrac{1}{4}\left(\frac{f_b}{f_m}\right)\left(\frac{c}{\Delta f}\right)$$

(13-28)

SINUSOIDAL MODULATION

For the average beat-frequency case, use the linear modulation equations.

FM CW FIXED ERROR

$$E_f = 246/\Delta f, \text{ feet}$$

(13-29)

Δf is in MHz.

DOPPLER FREQUENCY (f_d)

Direct Path

$$f_d = 2fV/c \qquad\qquad\qquad (13\text{-}30)$$

or

$$f_d = 2V/\lambda \qquad\qquad\qquad (13\text{-}31)$$

where

V is the relative velocity

c is the velocity of propagation ($3 \cdot 10^8$ meters per second)

f is frequency (in Hz)

λ is wavelength

Non-radial Path

The doppler frequency (f_d) between two points x and y is a function of the relative bearing angle θ as shown:

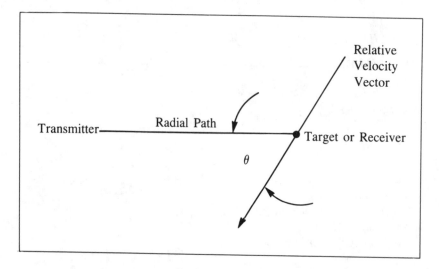

For Radar (Two-way Doppler)

$$f_d = 2\frac{v}{c}f_t \cos \theta \qquad\qquad\qquad (13\text{-}32)$$

For Communications (One-way Doppler)

$$f_d = \frac{v}{c}f_t \cos \theta \qquad\qquad\qquad (13\text{-}33)$$

v is velocity

f_t is the transmitter frequency

c is the propagation velocity
θ is the bearing angle

RECEIVING SYSTEMS

MIXERS

Mixer performance is described by

$$F_{if} = |Mf_l \pm Nf_r| \tag{13-34}$$

where

F_{if} is the intermediate frequency
f_l is the local-oscillator frequency
f_r is the received frequency
M and N are integers

Of the many possible combinations, minimum conversion loss results when both M and N equal 1. All other combinations result in greater loss. See Table 13-3.

EXAMPLE:

$$Nf_r = f_r \quad (N = 1)$$
$$Mf_l = f_l \quad (M = 1)$$

The conversion loss is shown as 0 dB. The actual value must include conversion loss of 6 to 7 dB added to the table values.

If $M = 3$, $N = 4$, the received signal power is 0 dBm for the M1D mixer, and the local oscillator power is 17 dBm, then the response is -80 dB relative to the value where $M = N = 1$. Note the value of the balancing as shown for even integers of M and N.

There are two responses for the $M = N = 1$ case resulting for $f_l > f_r$ and $f_r > f_l$. One is selected as the desired and the other an image response.

Spurious frequencies can be determined by rearranging equation 13-34 as follows and redefining f_r (the received frequency) to f_s (the spurious frequency).

$$f_s = \left| \frac{F_{if} - Mf_l}{\pm N} \right|$$

Images Resulting from Mixing

There are two responses resulting from the mixing process described by

$$F_{if} = |NF_r \pm Mf_{lo}| \tag{13-35}$$

where

F_{if} is the output of the mixing process and is the intermediate frequency
f_r is the received frequency
f_{lo} is the local oscillator frequency
M and N are integers

Table 13-3. Mixer Spurious Response Table for High-Performance Double-Balanced Mixers. (Courtesy Watkins-Johnson Company)

Typ. M1 | M1D | M1E

Harmonics of f_R

f_R \ f_L	0	1	2	3	4	5	6	7	8
7	79 / >99 >90 / >90	69 / >99 >90 / >90	80 / >99 >90 / >90	74 / >99 >90 / >90	78 / >99 >90 / >90	83 / >99 >90 / >90	63 / 78 >99 87 / >90	78 / >99 >90 / >90	60 81 / >99 >90 / >90 ; 71 99 / >99
6	90 / >99 >90 / >90	86 / >99 >90 / >90	91 / >99 >90 / >90	91 / >99 >90 / >90	90 / >99 >90 / >90	84 / >99 >90 / >90	84 / >99 >90 / >90	93 / >99 >90 / >90	84 / >99 97 ; 88 / >99 98
5	72 / 93 >99 >90 / >90	70 / 73 >90 / >90	71 / 87 >90 / >90	52 / 72 95 >90 / >90	77 / 88 >99 >90 / >90	46 / 66 >99 >90 / >90	75 / 85 >99 >90 / >90	45 / 64 90 >90 / >90	73 82 / >99 >90 / >90
4	80 / 96 88 86 / >90	79 / 80 >90 / >90	82 / 96 >99 86 / >90	77 / 80 92 >90 / >90	82 / 95 >90 / >90	76 / 82 95 85 / >90	85 / 98 86 / >90	72 / 78 94 85 / >90	77 / 90 87
3	51 / 81 67 / >90	49 / 58 64 / 77	53 / 65 85 69 / 87	51 / 60 69 50 / 78	55 / 65 85 77 / >90	48 / 55 68 47 / 75	54 / 64 85 54 / 74	53 / 54 64 44 / 77	58 / 66 87 ; 74 / 88
2	69 / 68 64 72 / 73	72 / 67 73 / 75	79 / 76 62 74 / 84	67 / 67 70 70 / 75	80 / 70 86 / 79	66 / 66 70 64 / 74	77 / 82 61 69 / 87	68 / 66 62 64 / 74	75 / 83 64 ; 69 / 84 79
1	25 / 25 24 24 / 23	0 / 0 0 / 0	39 / 39 35 35 / 39	13 / 11 13 / 11	45 / 50 42 40 / 46	22 / 16 19 24 / 14	54 / 59 50 45 / 62	37 / 19 39 28 / 19	59 / 49 49 ; 53 / 49
0	▨ (blank) 24 / 24	26 / 27 36 / 39	35 / 31 45 / 42	36 / 23 39 / 52	50 / 47 58 63 / 58	41 / 36 53 / 51	53 / 51 60 / 65	49 / 37 21 71 / 49	63 / 75 29 ; 51 / 63 19

Harmonics of f_L: 0   1   2   3   4   5   6   7   8

Intermodulation signals which result from the mixing of mixer-generated harmonics of the input signals are shown above. Mixing products are indicated by the number of dB below the $f_L \pm f_R$ output level.

The performance was measured under the following input conditions.

▨ (shaded) M1/M1D/M1E: f_R at 0 dBm; f_L at +7/+17/+27 dBm

☐ (unshaded) M1/M1D/M1E: f_R at −10 dBm; f_L at +7/+17/+27 dBm

f_R at 49 MHz, f_L at 50 MHz

Improved performance can be obtained at lower frequency, and f_R at a lower level.

Note: The mixer fundamental output is usually the desired output ($M = N = 1$) except for the harmonic mixing case where $M > 1$. All other outputs are considered to be spurious products.

The image is the response other than that desired for the case where M and $N = 1$ and is of the opposite sign ($\pm$).

INTERMODULATION DISTORTION (*IM*)

Given the two signals $A_1 \cos \omega_1 t$ and $A_2 \cos \omega_2 t$ across a nonlinear device, intermodulation produces result. The most significant output is   (13-36)

$$K_1[A_1 \cos \omega_1 t + A_2 \cos (\omega_2 t + \phi)] \tag{13-37}$$

Second-Order Distortion Terms   (13-38)

$$0.5K_2A_1^2 \cos 2\omega_1 t \tag{13-38}$$
$$0.5K_2A_2^2 \cos (2\omega_2 t + \phi) \tag{13-39}$$
$$K_2A_1A_2 \cos [\omega_1 t - \omega_2(t + \phi)] \tag{13-40}$$
$$K_2A_1A_2 \cos [\omega_1 t + \omega_2(t + \phi)] \tag{13-41}$$

Third-Order Terms

$$0.75K_3A_1^2A_2 \cos [2\omega_1 t \pm \omega_2(t + \phi)] \tag{13-42}$$
$$0.75K_3A_1A_2^2 \cos [2\omega_2(t + \phi) \pm \omega_1 t] \tag{13-43}$$

Cascade Intercept Point

Third Order

$$\overset{3}{I_t} = \overset{3}{I_2} - 10 \log \left[1 + \frac{1}{G_2} \cdot \frac{\overset{3}{I_2}}{\overset{3}{I_1}} \right] \tag{13-44}$$

where

$\overset{3}{I_t}$ is the third-order output intercept point total

$\overset{3}{I_2}$ is the second-stage output intercept point

G_2 is the second-stage gain

$\overset{3}{I_1}$ is the first-stage output intercept point

Second Order

$$\overset{2}{I_t} = \overset{2}{I_2} - 20 \log \left[1 + \left(\frac{1}{G_2} \frac{\overset{2}{I_2}}{\overset{2}{I_1}} \right)^{\frac{1}{2}} \right] \tag{13-45}$$

where

$\overset{2}{I_1}$ is the first stage output intercept point

$\overset{2}{I_2}$ is the second stage output intercept point

$\overset{2}{I_t}$ is the cascade output intercept point

G_2 is the second stage gain

The calculation for second or third order begins at the last stage in a configuration; count this as the second stage. The stage preceding this is considered the first stage. When that calculation is completed, again count the result as the second stage and the stage ahead of that is the first stage, etc.

The input intercept point is the output intercept point minus the intervening gain.

Normalizing Two Signals of Unequal Power

$$P_n \cong P_{\text{larger}} - \tfrac{1}{3}(P_{\text{larger}} - P_{\text{smaller}}) \tag{13-46}$$

All powers are in dB notation (dBm, dBW, etc.).

Intercept Point Given Distortion

$$\overset{m}{I} = S + (\overset{m}{R}/(m-1)) \tag{13-47}$$

m is the order of the distortion

$\overset{m}{I}$ is the m-order intercept point (output)

$\overset{m}{R}$ is the intermodulation ratio (in *dB*)

S is the signal strength of the two signals (in *dBm*)

SPURIOUS FREE DYNAMIC RANGE

$$\text{SFDR} = 0.67(\overset{3}{I} - kT/\text{MHz} - 10 \log_{10} B - NF), \text{ dB} \tag{13-48}$$

where

kT is -144 dBm per MHz

B is bandwidth (in *MHz*)

NF is the noise figure of the network or configuration

$\overset{3}{I}$ is the system's third-order input intercept point (output intercept $-$ gain $=$ input intercept)

S UNITS

The S unit is used to indicate the strength of a received signal. It has been generally accepted that a value of S9 is equal to a signal strength of 50 μV.

Values over S9 are in dB. Values below S9 are lower by 6 dB for each integrally decreasing S unit.

RADIO HORIZION

$$d = (3kh/2)^{\frac{1}{2}} \tag{13-49}$$

$k = \frac{4}{3}$ refraction correction
h is the antenna height (in *feet*)
$d = 1.414\ h^{\frac{1}{2}}$ (in *statute miles*) (13-50)
$d = 1.23\ h^{\frac{1}{2}}$ (in *nautical miles*) (13-51)

FRESNEL INTERFERENCE ZONES

Consider a plane that is orthogonal to the line of sight between a transmitting antenna and a receiving antenna. If it is made to contain concentric circles whose radii are such that a line from the receiver to a point on the circumference and then to the transmitter is half a wavelength longer than the line-of-sight distance, this circular zone is the first Fresnel zone. Similarly, other zones (nth zone) are $(n/2)\lambda$ for $n > 1$.

The first Fresnel zone includes approximately 25 percent of the total power. This radius is

$$R = 13.15[(\lambda D_1 D_2)/D]^{\frac{1}{2}} \quad \text{(in *feet*)} \tag{13-52}$$

where

λ is wavelength (in *cm*)
D_1 is the point on the Fresnel circle to transmitter (distance in *miles*)
D_2 is the point on the Fresnel circle to receiver (distance in *miles*)
D is the total line-of-sight distance between the transmitter and the receiver.

or

$$R = 2280[(D_1 D_2)/(DF)]^{\frac{1}{2}} \quad \text{(in *feet*)} \tag{13-53}$$

F is frequency in MHz.

When $D_1 = D_2$,

$$R = 1140(D/F)^{\frac{1}{2}} \quad \text{(in *feet*)} \tag{13-54}$$

FREQUENCY AND WAVELENGTH

$$\lambda = \frac{c}{F} \tag{13-55}$$

where

λ is wavelength
c is the velocity of propagation
F is frequency

$$\lambda \text{ in meters} = \frac{300{,}000}{F(\text{kHz})} = \frac{300}{F(\text{MHz})} \tag{13-56}$$

$$\lambda \text{ in feet} = \frac{984}{F(\text{MHz})} \tag{13-57}$$

14

---
modulation

AMPLITUDE MODULATION

SINUSOIDAL CASE

$$A(1 + m \cos \mu t) \cdot \sin (\omega t + \theta) \qquad (14\text{-}1)$$

where

 A is the magnitude of the carrier
 m is the modulation index or factor $(0 < m < 1.0$, and $m = 1.0 = 100\%)$
 $\mu = 2\pi f$
 f is the modulation frequency
 $\omega = 2\pi F$
 F is the carrier frequency
 θ is a phase angle

Expanding gives

$$A \sin (\omega t + \theta) + Am/2\{\sin [(\omega + \mu)t + \theta]\} \\ + Am/2\{\sin [(\omega - \mu)t + \theta]\} \qquad (14\text{-}2)$$

This expansion shows that the AM signal consists of the carrier and two sidebands.

COMPLEX MODULATION CASE

The expression $g(t)$ is the complex modulation represented as a series. Then

$$A[1 + mg(t)] \cdot \sin (\omega t + \theta) \qquad (14\text{-}3)$$

AM Spectral Energy

Total for the Sinusoidal Case

$$A^2\left(1 + \frac{m^2}{2}\right)$$

(14-4)

where

A is magnitude

m is the modulation factor

Components:

The carrier is A^2

The lower sideband $\left(\dfrac{Am}{2}\right)^2$

The upper sideband $\left(\dfrac{Am}{2}\right)^2$

Modulation Percentage

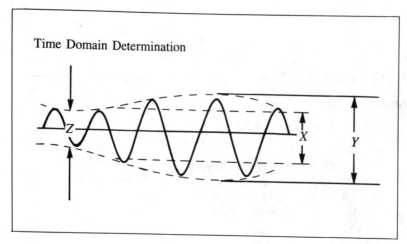

Time Domain Determination

Upward Modulation

$$m\% = 100\left(\frac{Y - X}{X}\right)$$

(14-5)

Downward Modulation

$$m\% = 100\left(\frac{X - Z}{X}\right)$$

(14-6)

Amplitude Modulation Signal-to-Noise Ratio

$$S/N = \frac{P_c m^2}{2kTBF}$$

(14-7)

where

S/N is the signal-to-noise ratio

P_c is the carrier power

m is the modulation factor ($m = 1 = 100\%$; $m = 0.5 = 50\%$)

k is $1.38 \cdot 10^{-23}$ joules per degree Kelvin

T is 290 degrees Kelvin

F is noise factor

B is the post-detection bandwidth

In dB notation:

$$S/N = P_c + m^2 - 3 - kT - B - NF \tag{14-8}$$

AMPLITUDE SHIFT KEYING, CW TELEGRAPHY, AND ON/OFF KEYING

RF Carrier Time Domain Representation (Morse A in CW Telegraphy)

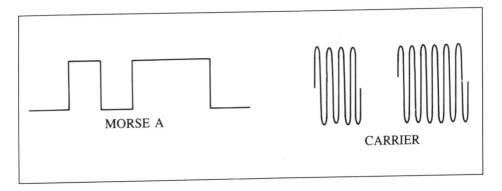

MORSE A

CARRIER

Detection Error Probability

Noncoherent or Envelope Detection

For minimum probability of error, the detection threshold must be

$$(A/2)\left(1 + 2\frac{e_b}{N_o}\right)^{\frac{1}{2}} \tag{14-9}$$

where

e_b is pulse energy

N_o is noise density

A is the pulse amplitude

Given the decision threshold equals $A/2$,

$e_b/N_o \gg 1$

then the probability of error for mark is:

$$P_{e_{mark}} = \frac{1}{(2\pi e_b/N_o)^{\frac{1}{2}}} e^{(-e_b/2N_o)}$$

(14-10)

For Space:

$$P_{e_{space}} = e^{(-e_b/2N_o)}$$

(14-11)

Coherent Amplitude Shift Keying Detection Probability of Error

$$P_e = \frac{1}{2} erfc\left(\frac{e_b}{2N_o}\right)^{\frac{1}{2}}$$

(14-12)

The expression *erfc* is the complementary error function. There is more information in Chapter 21 under the Error Function and Complementary Error Function subheadings.

PULSE MODULATION

SPECTRUM

$$A_j = 2A\frac{T}{t_r}\left[\frac{\sin \pi Tj/t_r}{\pi Tj/t_r}\right]$$

(14-13)

where

  A is amplitude
  T is the pulse width
  t_r is the repetition period
  j is the harmonic of $f_r = 1/t_r$

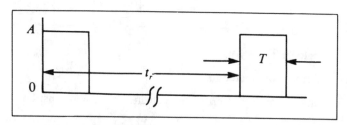

PULSE ENVELOPE IN THE TIME DOMAIN

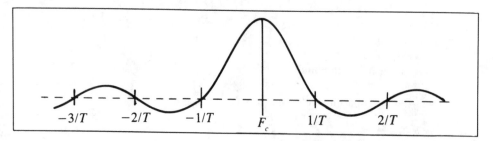

SPECTRAL ENVELOPE IN THE FREQUENCY DOMAIN

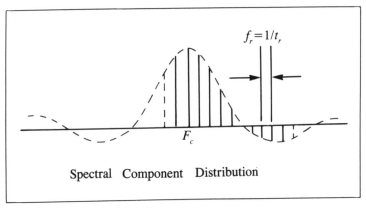

Spectral Component Distribution

ALTERNATE MARK INVERSE

The principle use of this form of encoding is used in low-bandwidth systems.

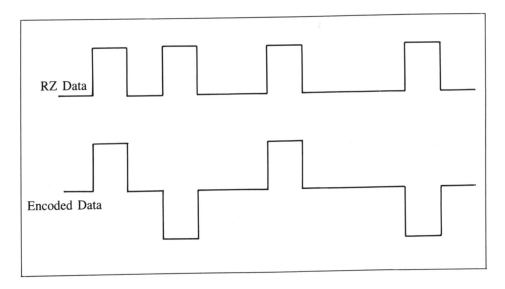

AM SINGLE-SIDEBAND SUPPRESSED CARRIER_____

This signal is a more efficient form of AM where only one of the sideband sets of an AM signal is transmitted. The other sideband set and the carrier are removed by filtering or phasing.

FOR UPPER-SIDEBAND TRANSMISSION

$$\frac{Am}{2} \{\sin\left[(\omega + \mu)t + \theta\right]\} \tag{14-14}$$

FOR LOWER-SIDEBAND TRANSMISSION

$$\frac{Am}{2} \{\sin [(\omega - \mu)t + \theta]\} \tag{14-15}$$

Term definitions follow equation (14-1).

AM DOUBLE-SIDEBAND SUPPRESSED CARRIER

This signal is generated by removing the carrier from the AM signal.

$$\frac{Am}{2} [\sin (\omega + \mu)t + \sin (\omega - \mu)t + \theta] \tag{14-16}$$

Term definitions follow equation (14-1).

DETECTOR CARRIER REINSERTION

Inserted carrier frequency accuracy at the product detector must be 20 to 80 Hz or less.

ANGULAR MODULATION

MODULATION INDEX

Phase Modulation (PM) Index

$\Delta\theta$ peak is the mod index for PM (in *radians*)

θ is phase deviation of the carrier caused by modulating signal

Frequency Modulation (FM) Index

$$\frac{\Delta f \text{ peak}}{\mu} \tag{14-17}$$

f is the carrier frequency deviation (in *Hz*)

μ is the modulation frequency signal (in *Hz*)

Let

$\Delta\theta$ peak $= \beta = \Delta f$ peak$/\mu$

β is the argument of the Bessel function of the first kind of order n where n is an integer and described by $J_0(\beta)$, $J_1(\beta)$ $J_2(\beta)$ . . . $J_n(\beta)$. The value of these Bessel functions is representative of the magnitude of the signal's sidebands plus and minus n.

To determine the spectra of an FM or PM signal, the values of $J_n(\beta)$ can be read from tables or charts of Bessel functions. For most applications, chart readouts are adequate.

EXAMPLE:

For $\beta = 5$,

n	$J_n(\beta)$	
0	0.18	Carrier
±1	0.32	
±2	0.05	
±3	0.38	
±4	0.39	Sidebands
±5	0.26	
±6	0.15	
±7	0.05	

The absolute value of $J_n(\beta)$ is the magnitude seen on an spectrum analyzer. Sidebands are displaced to either side of the carrier by $\pm n$. A spectrum analyzer responsive to power will display $[J_n(\beta)]^2$ normalized to the largest term ($n = 4$), as shown in the following table:

n	Relative Power
0	0.213
±1	0.673
±2	0.0164
±3	0.95
±4	1.0
±5	0.44
±6	0.111
±7	0.0164

As an example, the graph on p. 188 shows the power spectrum of an FM wave for a β of 5.

FM SPECTRUM

For systems with an index (β) less than 20

Systems adjustments of deviation can be made by choosing a ratio of ΔF and μ that dictates the magnitude of a certain selected sideband as seen on a spectrum analyzer.

EXAMPLE:

Desired deviation = 24 kHz
Modulating frequency range = 300 to 3000 Hz.

The range of $\beta = \Delta F/\mu$ is computed to be $8 < \beta < 80$. From the Bessel curves below the 2nd maxima of the third sideband as β increases from 0 is a β of 8. From

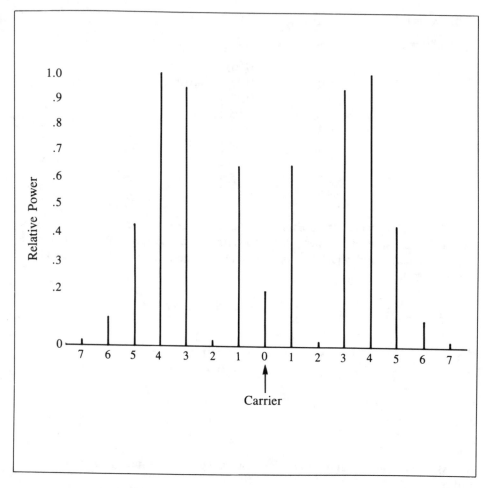

this, the modulating frequency must be 3000 Hz and the magnitude of the modulation should be increased until the second maxima of the third sideband is achieved. This is the desired level of the modulation.

The magnitude of the spectral components for any low value of β (less than 11.5) can be read directly from the Bessel curves. For higher values of β (11.5 < β < 20), the high-order sideband magnitude can be determined. Note that as an example, the 13th sideband is the 13th harmonic of the modulating frequency.

FREQUENCY MODULATION (FM)

SINUSOIDAL CASE

$$A \sin \left[2\pi Ft + \frac{\Delta F}{f} \sin 2\pi ft + \theta \right]$$

(14-18)

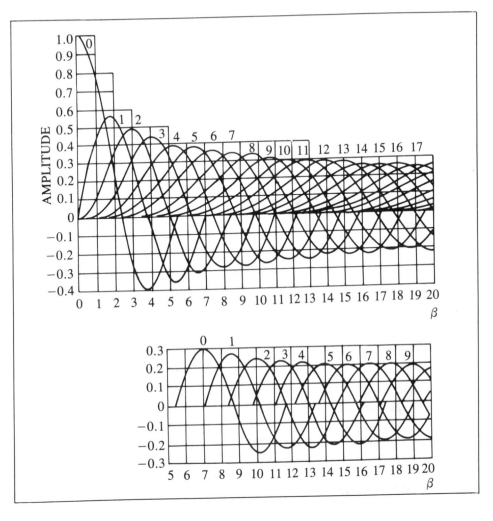

When $\Delta F \gg f$

$$A \sin (\omega t + \Delta F \sin \mu t + \theta) \qquad (14\text{-}19)$$

where

    A is magnitude

    F is the carrier frequency

    f is the modulating frequency

    θ is phase

    ΔF is the peak deviation of the carrier

    ω is $2\pi F$

    μ is $2\pi f$

    $\Delta F/f$ is the modulation index

    t is time

Spectral Components

$$A\{J_o(\beta) \sin(\omega t + \theta) + J_n(\beta)[\sin(\omega t + n\,\mu t) \pm \sin(\omega t - n\,\mu t)]\} \qquad (14\text{-}20)$$

The sign ($\pm$) is ($+$) when n is even and ($-$) when n is odd.

FM Bandwidth (or Carson's Bandwidth)

Carson's Rule

$$B_{IF} \approx 2(\Delta F + 2f_m) \qquad (14\text{-}21)$$

where

B_{IF} is the IF bandwidth
ΔF is the peak deviation of the carrier
f_m is the modulating frequency
$\Delta F/f_m = B$
$2 < B < 10$

FM Bandwidth Determination by Spectral Power Summation

The sum of the square of the sideband Bessel coefficients must equal unity. From this, a bandwidth can be selected to accommodate or include spectral lines for a particular efficiency.

$$[J_o(B)]^2 + 2[J_1(B)]^2 + 2[J_2(B)]^2 + \ldots 2[J_n(B)]^2 = 1 \qquad (14\text{-}22)$$

where

J_n is a Bessel function of the first kind
n is the order $(0, 1, 2, 3, 4, 5, \ldots n)$
B is the modulation index $= \Delta F/f$
ΔF is the peak carrier deviation
f is the modulating frequency
$J_o(B)$ is the carrier component magnitude
$J_1(B)$ is the magnitude of the first sideband terms
$J_2(B)$ is the magnitude of the second sidebands.
$J_n(B)$ is the magnitude of the nth sidebands and is a Bessel function of order n and argument B. (Bessel functions can be found in *Tables of Higher Functions* by E.F. Jahnke and F. Lösch. See Bibliography.)

FM Carrier-to-Noise Ratio (P_c/N)

$$P_c/N = P_c/(kTBF) \qquad (14\text{-}23)$$

where

P_c is the carrier power
k is $1.23 \cdot 10^{-23}$ joules per degree Kelvin
T is 290 degrees Kelvin
B is the effective noise bandwidth
F is noise factor
N is the noise power

In dB notation:

$$(P_c/N)_{dB} = P_c - k - T - B - NF \tag{14-24}$$

where

NF is $10 \cdot \log_{10} (F)$ (in *decibels*)

FM OUTPUT SIGNAL-TO-NOISE RATIO

For S/N above the threshold of 10 dB,

$$(S/N)_{out} = (P_c/N)(MNI) \tag{14-25}$$

where

$(S/N)_{out}$ is the output signal-to-noise ratio
P_c/N is the carrier-to-noise ratio at the limiter
MNI is the modulation noise improvement factor (see next section)

MODULATION NOISE IMPROVEMENT FACTOR

$$MNI = \left(\frac{3}{2}\right)\left(\frac{\Delta F}{B_o}\right)^2\left(\frac{B_{IF}}{B_o}\right) \tag{14-26}$$

ΔF is the peak deviation of the carrier
B_o is the audio- or output-noise-bandwidth
B_{IF} is the IF bandwidth

The carrier-to-noise ratio must be greater than 10 dB.

FM OUTPUT SIGNAL-TO-NOISE RATIO BELOW THRESHOLD

For a carrier-to-noise ratio below 10:

$$S/N_{out} = MNI\left(\frac{P_c}{N}\right)\left[\frac{1}{1 + 0.9\left(\frac{B_{IF}}{B_o}\right)^2 \frac{(P_c/N)e^{-(P_c/N)}}{[1 - e^{-(P_c/N)}]^2}}\right] \tag{14-27}$$

While $e = 2.718$, the other term definitions follow equation (14-25) and (14-26).

FREQUENCY-SHIFT KEYING (FSK)

ERROR PROBABILITY (P_e)

Noncoherent Detection

$$P_e = \frac{1}{2}\exp\left[-\frac{1}{2}\frac{e_b}{N_o}\right] \tag{14-28}$$

where

e_b is energy per bit
N_o is noise density

Coherent Detection

$$P_e = \frac{1}{2} erfc \left[\frac{1}{2} \frac{e_b}{N_o} \right]^{\frac{1}{2}}$$

(14-29)

where

erfc is the complementary error function

FSK SPECTRAL ENVELOPE

$$F_\omega = \frac{1}{2} AT \left[e^{jb} \frac{\sin x}{x} + e^{-jb} \frac{\sin y}{y} \right]$$

(14-30)

where

$x = \frac{1}{2} T(\omega - \omega_o)$

$y = \frac{1}{2} T(\omega + \omega_o)$

ω is $2\pi f_{carrier}$

ω_o is $2\pi f_{shift}$

A is a magnitude

T is the frequency shift time duration

e is 2.718

j is $\sqrt{-1}$

b is $(\theta - \pi/2)$

$e^{jb} = \cos b + j \sin b$

$e^{-jb} = \cos b - j \sin b$

The spectrum is identical in envelope to that of a repetitive pulse except for its duality. There are two such spectral envelopes for the FSK signal. The spacing is determined by the mark space frequency differential.

MARK AND SPACE FILTER BANDWIDTH

A bandwidth for each filter of $1.5f_m$ is a typical choice for good performance (f_m is the data rate).

RF AND IF FILTER-BANDWIDTH

$$B_{\text{IF and RF}} \cong 2(f_m + S)$$

(14-31)

where

f_m is $1/T$ or the data rate

S is the mark space frequency spacing

PHASE MODULATION

SINUSOIDAL CASE

Mathematical representation of the spectrum is

$$A \sin (\omega t + \Delta\theta \cos \mu t + \phi)$$

(14-32)

where
  A is magnitude
  ω is $2\pi F$
  F is carrier frequency
  t is time
  $\Delta\theta$ is the peak phase deviation of the carrier
  μ is the modulating signal ($2\pi f$)

SPECTRAL COMPONENTS

$$
\begin{aligned}
A\,[& J_0(\Delta\theta)\,\sin\,(\omega t + \phi) \\
& + J_1(\Delta\theta)\,\cos\,(\omega t + \mu t + \phi) + J_1(\Delta\theta)\,\cos\,(\omega t - \mu t + \phi) \\
& - J_2(\Delta\theta)\,\sin\,(\omega t + 2\mu t + \phi) - J_2(\Delta\theta)\,\sin\,(\omega t - 2\mu t + \phi) \\
& - J_3(\Delta\theta)\,\cos\,(\omega t + 3\mu t + \phi) - J_3(\Delta\theta)\,\cos\,(\omega t - 3\mu t + \phi) \\
& + J_4(\Delta\theta)\,\sin\,(\omega t + 4\mu t + \phi) + J_4(\Delta\theta)\,\sin\,(\omega t - 4\mu t + \phi) \\
& + J_5(\Delta\theta)\,\cos\,(\omega t + 5\mu t + \phi) \, . \, . \, . \,]
\end{aligned}
\tag{14-33}
$$

BANDWIDTH REQUIREMENTS FOR PHASE MODULATION

Carson's Rule

$$
B_{IF} = 2(M + 2)f_m \tag{14-34}
$$

where
  B_{IF} is the IF bandwidth
  M is the modulation index for phase modulation
  f_m is the modulating frequency (highest)
  $2 < M < 10$

BANDWIDTH DETERMINATION BY POWER SUMMATION

The signal energy is proportional to the square of the Bessel coefficients included in the summation. The required bandwidth can be determined for a particular efficiency.

$$
J_o(M)^2 + 2[J_1(M)]^2 + 2[J_2(M)]^2 + 2[J_3(M)]^2 + \; . \, . \, . \; 2[J_n(M)]^2 = 1 \tag{14-35}
$$

where
  $J_n(M)$ is a Bessel function of the first kind
  M is the phase modulation index is $\Delta\theta$, which is the carrier frequency phase deviation (peak), in *radians*
  n is the order of the Bessel function
  M is the argument of the Bessel function

PHASE MODULATION SIGNAL-TO-NOISE RATIO

Above Threshold (Where $S/N > 10$)

$$
(S/N)_{out} = M^2\left(\frac{B_{IF}}{2B_o}\right)\frac{P_c}{N} \tag{14-36}
$$

where

$(S/N)_{out}$ is the output signal-to-noise ratio

M is the peak phase deviation

B_{IF} is the IF equivalent noise bandwidth (ENB)

B_0 is the output audio bandwidth

N is the noise in the IF bandwidth

P_c is the carrier power

Below Threshold (Where $S/N < 10$)

$$(S/N)_{out} = M^2\left(\frac{B_{IF}}{2B_o}\right)\left(\frac{P_c}{N}\right)\left[\frac{1}{1 + 0.9\left(\frac{B_{IF}}{B_o}\right)^2 \dfrac{(P_c/N)e^{-P_c/N}}{(1 - e^{-P_c/N})^2}}\right] \qquad (14\text{-}37)$$

Variable definitions follow equation (14-36).

PHASE-SHIFT KEYING

SPECTRUM

$$F_\omega = AT\left[\frac{\sin{(\omega_1 - \omega_0)}T/2 + \phi}{(\omega_1 - \omega_0)T/2 + \phi}\right] \qquad (14\text{-}38)$$

where

A is amplitude

T is the bit width

ω_1 is the carrier radian frequency for a 1

ω_0 is the carrier radian frequency for a 0

ϕ is phase

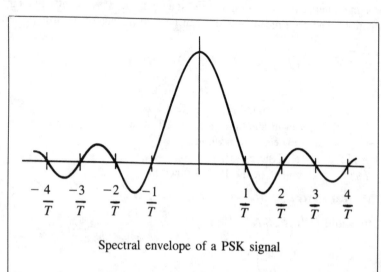

Spectral envelope of a PSK signal

PHASE-SHIFT KEYING ERROR PROBABILITY (P_e)

Coherent Detection

$$P_e = \frac{1}{2} erfc \sqrt{\frac{e_b}{N_o}}$$

(14-39)

Signal Reference Derived Detection

$$P_e = \tfrac{1}{2} e^{-(e_b/N_o)}$$

(14-40)

erfc is the complementary error function
e_b/N_o is the energy per bit to noise density
e is 2.178

At low e_b/N_o coherent detection has a 3 dB advantage over the signal-derived reference detection method. At strong signal conditions, there is approximately a 1 dB difference between both systems.

15

phase-locked loops

BASIC CONFIGURATION

PHASE DETECTOR

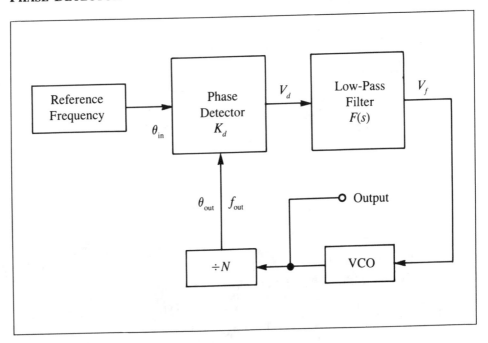

The phase detector output (V_d) is a DC potential proportional to the phase difference $\Delta\phi$ between the reference and VCO (f_o) frequency.

$$\Delta\phi = \phi_{in} - \phi_{out} \tag{15-1}$$

$$V_d = K_d \, \Delta\phi \quad (in \; volts) \tag{15-2}$$

where

K_d is the phase detector conversion gain (in *volts per radian*)

$\Delta\phi$ is the phase difference between the reference frequency and f_o (in *radians*)

VOLTAGE-CONTROLLED OSCILLATOR (VCO)

The VCO is an oscillator whose frequency (f_o) is controllable by an external DC potential V_f.

$$2\pi f_o = \omega_o = K_o V_f \quad (in \; radians \; per \; second) \tag{15-3}$$

K_o is the VCO sensitivity of frequency change to a change in tuning potential (in *radians per second per volt*)

V_f is the DC tuning potential (in *volts*)

THE PHASE-LOCKED LOOP FILTER

The low-pass filter transfer function is represented by $F(s)$ (Laplace notation) such that its behavior can be represented by

$$V_f(s) = V_d F(s) \tag{15-4}$$

$$= K_d(\phi_{in} - \phi_{out})F(s) \tag{15-5}$$

$$= \Delta\phi K_d F(s) \tag{15-6}$$

The three filter forms that are in popular use follow.

Passive RC Low-Pass Filter

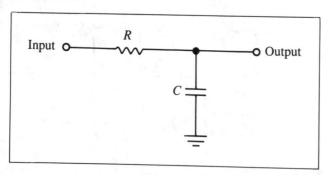

$$F(s) = 1/(1 + sRC) \tag{15-7}$$

Passive Lead/Lag RC Low-Pass Filter

$$F(s) = \frac{1 + sCR_2}{1 + sC(R_1 + R_2)} = \frac{1 + sT_2}{1 + s(T_1 + T_2)} \tag{15-8}$$

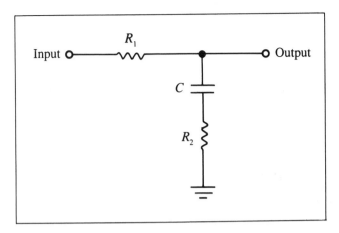

Active Lead/Lag Low-Pass Filter

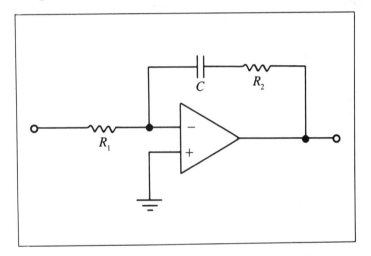

$$F(s) = \frac{1 + sCR_2}{sCR_1} = \frac{1 + sT_2}{sT_1} \qquad (15\text{-}9)$$

BASIC SECOND-ORDER LOOP EQUATIONS

LOOP TRANSFER FUNCTION ($H(s)$)

$$H(s) = \frac{\phi_{\text{out}}}{\phi_{\text{in}}} = \frac{K_o K_d F(s)}{s + K_o K_d F(s)} \qquad (15\text{-}10)$$

where
   K_o is defined following equation (15-3)
   K_d is the phase detector conversion gain constant (in *volts per radian*)

Passive RC Low-Pass Filter

$$H(s) = \frac{K_o K_d/T}{s^2 + s(1/T) + K_o K_d/T} = \frac{\omega_n^2}{s^2 + 2\delta\omega_n s + \omega_n^2} \tag{15-11}$$

δ is the damping factor where

$$\delta = \frac{1}{2}\left[\frac{1}{K_o K_d T}\right]^{\frac{1}{2}} \tag{15-12}$$

ω_n is the natural loop frequency where

$$\omega_n = \left[\frac{K_o K_d}{T}\right]^{\frac{1}{2}} \tag{15-13}$$

Passive Lead/Lag RC Low-Pass Filter

$$H(s) = \frac{K_o K_d (sT2 + 1)/(T1 + T2)}{s^2 + \dfrac{s(1 + K_o K_d T2)}{T1 + T2} + \dfrac{K_o K_d}{T1 + T2}} \tag{15-14}$$

$$= \frac{s\omega_n[2\delta - \omega_n/(K_o K_d)] + \omega_n^2}{s^2 + 2\delta\omega_n s + \omega_n^2} \tag{15-15}$$

δ is the damping factor where

$$\delta = \frac{1}{2}\left[\frac{K_o K_d}{T1 + T2}\right]^{\frac{1}{2}}\left[T2 + \frac{1}{K_o K_d}\right] \tag{15-16}$$

ω_n is the natural frequency where

$$\omega_n = \left[\frac{K_o K_d}{T1 + T2}\right]^{\frac{1}{2}} \tag{15-17}$$

$T1 = R_1 C$, $T2 = R_2 C$

Active Lead/Lag Low-Pass Filter

$$H(s) = \frac{[K_o K_d (sT2 + 1)]/T1}{s^2 + s\dfrac{K_o K_d T2}{T1} + \dfrac{K_o K_d}{T1}} \tag{15-18}$$

$$H(s) = \frac{2\delta\omega_n s + \omega_n^2}{s^2 + 2\delta\omega_n s + \omega_n^2} \tag{15-19}$$

where
$T1 = R_1 C$, $T2 = R_2 C$

δ is the damping factor where

$$\delta = \tfrac{1}{2}T2\left[\frac{K_o K_d}{T1}\right]^{\frac{1}{2}} \tag{15-20}$$

ω_n is the natural frequency where

$$\omega_n = \left[\frac{K_o K_d}{T1} \right]^{\frac{1}{2}} \tag{15-21}$$

LOOP BANDWIDTH FOR PASSIVE LEAD/LAG FILTER

Let $|H(s)|^2 = \frac{1}{2}$. Solve for ω by substituting $s = j\omega$.

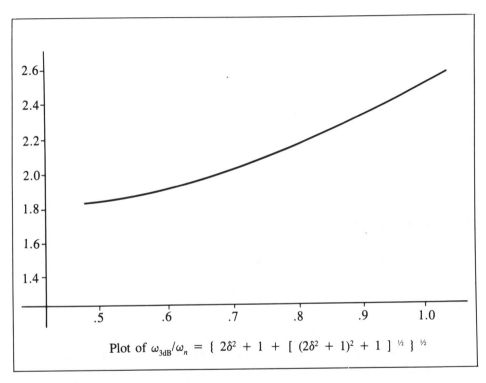

Plot of $\omega_{3dB}/\omega_n = \{ 2\delta^2 + 1 + [(2\delta^2 + 1)^2 + 1]^{\frac{1}{2}} \}^{\frac{1}{2}}$

ω_{3dB} is the 3 dB frequency
ω_n is the natural frequency [see equation (15-17)]
δ is the damping factor [see equation (15-16)]

DAMPING FACTOR (WHERE $\delta < 1$)

Introduce a step input into the loop and observe the loop filter output.

$$y' = \frac{1}{2\pi} \ln \left[\frac{y_1}{y_2} \right] \tag{15-22}$$

and

$$\delta = \frac{y'}{[1 + (y')^2]^{\frac{1}{2}}} \tag{15-23}$$

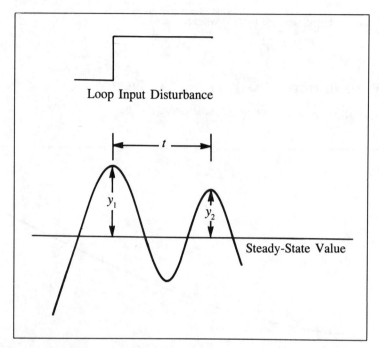

Loop Input Disturbance

Steady-State Value

DAMPING FREQUENCY (ω_d AND NATURAL FREQUENCY (ω_n)

$$\omega_d = 2\pi\frac{1}{t} \tag{15-24}$$

$$= \omega_n(1 - \delta^2)^{\frac{1}{2}}$$

$$\omega_n = \omega_d\frac{1}{(1 - \delta^2)^{\frac{1}{2}}} \tag{15-25}$$

t is the time between successive peaks of the loop filter output voltage resulting from a loop step disturbance (see Damping Factor)

16

satellites

IDEAL ELLIPTIC ORBIT

Mass of the central body is very much greater than that of the satellite.

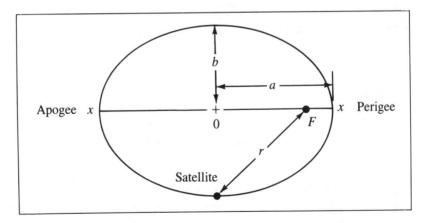

DISTANCE FROM THE CENTRAL BODY (FROM ZERO TO F, THE FOCUS)

$$0F = (a^2 - b^2)^{\frac{1}{2}} \qquad (16\text{-}1)$$

where
$$a > b$$

ORBITAL PERIOD (T)

$$T = \frac{2\pi a^{\frac{3}{2}}}{\mu^{\frac{1}{2}}} \quad \text{(in } hours\text{)} \tag{16-2}$$

$\mu = 5.164 \cdot 10^{12} \text{ km}^3/\text{hr}^2$
a is distance (in *kilometers*)

INSTANTANEOUS VELOCITY (v_i)

$$v_i = \left[\frac{\mu(2a - r)}{ra} \right]^{\frac{1}{2}} \tag{16-3}$$

ECCENTRICITY (e)

$$e = \frac{(a^2 - b^2)^{\frac{1}{2}}}{a} \tag{16-4}$$

r is the distance from the satellite to the focus
a is the semimajor axis or barycenter
b is the semiminor axis

THE SYNCHRONOUS OR STATIONARY ORBITAL CASE

The satellite position remains fixed relative to a point on the earth (the communications geometry).

ALTITUDE RELATIVE TO THE EARTH'S CENTER (h)

$$h = r_e + h_s \tag{16-5}$$

where

r_e = Earth's radius

Equatorial	6378.165 km
Polar	6356.785 km
Approximate	6378.280 km

h_s is the satellite height

The period of a synchronous satellite is one siderial day, which is equal to 23 hours, 56 minutes, and 4.009054 seconds. Therefore, solving for a in equation (16-2) yields

$$a = \left[\frac{T\mu^{\frac{1}{2}}}{2\pi} \right]^{\frac{2}{3}} = 42{,}162 \text{ km} \tag{16-6}$$

Thus, the computed height of the satellite from the center of the earth is 42,162 km. (The accepted value is 42,230 km). The earth radius is taken as 6300 km for a satellite height above earth of 35,930 km (35,930 + 6300 = 42,230).

17

sound

VELOCITY OF SOUND IN AIR

$$V = 1087(T/273)^{\frac{1}{2}} \tag{17-1}$$

where

V is velocity (in *feet per second*)
T is temperature (in *degrees Kelvin*)

SOUND ATTENUATION TABLE

This table shows the distance sound travels before being attenuated by one-half of its original value. This is shown as a function of frequency and distance in air or water.

Table 17-1.   Sound Travel Distance

Frequency (kHz)	Distance In Water (ft)	In Air (ft)
10	1,312,400	722.000
50	52,500	15.700
100	13,124	7.220
500	525	0.157
1000	131	0.072

VOLUME UNITS (VU)

Zero VU is equal to 1 milliwatt across 600 ohms.

$$\text{VU}_{dB} = 30 + 10 \log_{10} P \tag{17-2}$$

where
   P is power in watts across 600 ohms

MICROPHONE SENSITIVITY

Low impedance: dB as referred to 1 milliwatt for 10 dynes/cm^2.
High impedance and carbon: dB as referred to 1 volt for 1 dyne/cm^2.

Test Frequencies

   Voice applications—1000 Hz
   Music applications—250 Hz

Acoustic Pressure, Typical

The threshold of human hearing is 0.0002 dyne/cm^2 or 0dB.

Table 17-2.   Acoustic Pressures

Source	Distance	Peaks	Output (dB)	Dynes/cm^2
Quiet Studio	—	—	10	0.0006
Voice	$\frac{1}{4}$ in	+12 dB	107	44
15-piece				
Orchestra	10 ft	+20 dB	72	0.8
Pain Threshold	—	—	130	640

SONAR

ATTENUATION

$$A = 20 \log_{10} R + (40f^2/(4100 + f^2) + 2.75 \cdot 10^{-4}f^2)R \tag{17-3}$$

where
   A is the attenuation in dB
   R is range in kilo-yards
   f is frequency in kilo Hz

PASSIVE SONAR CASE
(Noise-limited)

$$A = n_t - D_t - n_b + I_d \tag{17-4}$$

where

n_t is target noise
n_b is background noise
D_t is the detection threshold differential (in dB)
I_d is directivity (in dB)

The terms n_b and n_t are referred to 1 μb/yd/Hz.

ACTIVE SONAR CASE

$$A = 0.5(P_s + I_d + P_t - n_b - B - D_t) \tag{17-5}$$

where

P_s is the source level (in dB as referred to 1 bar/yd)
P_t is the target level (in dB)
B is bandwidth (in dB as referred to 1 Hz)

Other terms are defined following equation (17-4).

SEA STATE

Table 17-3.   Sea State Number versus
Wave Height

Sea State	Wave Height (ft)	Number
Calm	0	0
Smooth	<1	1
Slight	1-3	2
Moderate	3-5	3
Rough	5-8	4
Very Rough	8-12	5
High	12-20	6

18

---

environmental

SHOCK DUE TO DECELERATION

CONSTANT-DECELERATION CASE

$$G_d = g(H/D) \qquad\qquad (18\text{-}1)$$

where
  G_d is the impact shock load
  H is the free-fall distance
  D is the stopping distance
  g is 32 feet per second per second

LINEAR DECELERATION CASE

$$G_l = 2g(H/D) \qquad\qquad (18\text{-}2)$$

VIBRATION AMPLITUDE (G)

$$G = 0.10225(df)^2 \qquad\qquad (18\text{-}3)$$

where
  G is the vibration amplitude (in *gravitational units*)
  d is the peak displacement (in *inches*)
  f is the frequency (in *Hertz*)

CENTRIFUGE

CENTRIFUGAL FORCE

$$F = \frac{Wv^2}{gr} \qquad (18\text{-}4)$$

where

W is the weight of the object
v is the velocity
g is 32 feet per second per second
r is the radius

VELOCITY (v)

$$v = \frac{(\text{circumference})(\text{rpm})}{60} \qquad (18\text{-}5)$$

$$= \frac{(2\pi r)(\text{rpm})}{60} \qquad (18\text{-}6)$$

Therefore, from above,

$$F = \frac{W}{gr}\left[\frac{(2\pi r)(\text{rpm})}{60}\right]^2 \qquad (18\text{-}7)$$

In gravitation units

$$G = \frac{F}{W} \qquad (18\text{-}8)$$

19

---

heat

FAN COOLING

The required flow is

$$\text{CFM} = 3160 \text{ kW}/\Delta T_F \qquad\qquad\qquad (19\text{-}1)$$

$$\text{CFM} = 1760 \text{ kW}/\Delta T_C \qquad\qquad\qquad (19\text{-}2)$$

where

CFM is cubic feet per minute
kW is kilowatts to be removed
ΔT is the allowable thermal rise (in *degrees*)
F is Fahrenheit
C is Centigrade

FAN PRESSURE HEAD

$$E_p = I_{\text{cfm}} R_s \qquad \text{(in *inches of water*)} \qquad\qquad (19\text{-}3)$$

where

E_p is the fan pressure head
I_{cfm} is the incoming air flow ratio in cubic feet per minute through the system
R_s is the air flow resistance in inches of water per cubic feet per minute

AIR STREAM TEMPERATURE RISE (V_r)

$$V_r = R_a I \qquad\qquad\qquad (19\text{-}4)$$

where

R_a is the airstream thermal resistance at the flow rate (in *degrees C per watt*)

I is the equipment heat transfer to air stream (in *watts*)

Also,

$$R_a = 1.76 \text{ /cfm} \tag{19-5}$$

or

$$V_r = 1.76 \text{ } I\text{/cfm} \tag{19-6}$$

RADIATION HEAT LOSS

RADIATED ENERGY BY A BLACK BODY: STEFAN BOLTZMANN LAW

A black body is a body that absorbs all incident radiation. It is a perfect radiator. A perfect radiator is a perfect absorber.

$$Q = 1.73 \cdot 10^{-9} e[(t_1 + 460)^4 - (t_2 + 460)^4] \tag{19-7}$$

where

Q is Btu per hour per square foot

t_1 is the temperature of the device (in *degrees F*)

t_2 is the temperature of the surrounding air (in *degrees F*)

e is the emissivity (or emittance) of the black body ($0 < e \leq 1$). It is a ratio of the rate of radiant emission from a body (resulting from its temperature) to that of the radiant emission from a black body of the same temperature.

EXAMPLE:

The emissivity of a surface is 0.9 at a temperature of 250°F. The surrounding air is at 75°F. Find the radiated energy.

$$Q = 1.73 \cdot 10^{-9} e[(250 + 460)^4 - (75 + 460)^4]$$
$$= 268 \text{ Btu/hr/sq ft}$$

HEATSINKING OF TRANSISTORS

JUNCTION TEMPERATURE (T_j)

$$T_j = P_d \theta_{ja} + T_a \tag{19-8}$$

where

P_d is power dissipation

$$\theta_{ja} = \theta_{jc} + \theta_{cs} + \theta_{sa} \tag{19-9}$$

θ_{jc} is transistor junction-to-case thermal resistance

θ_{cs} is insulator thermal resistance (case to sink)

θ_{sa} is heatsink thermal resistance (heatsink to ambient)

θ_{ja} is the total thermal resistance (junction to ambient)

T_a is the ambient temperature

Units

P_d	W	power dissipation
θ	°C	thermal resistance
C	Ws/°C	capacitance thermal
ΔT	°C	thermal rise

HEATSINK DEFINITION

$$\theta_{sa} = \frac{T_{j\max} - T_a}{P_d} - \theta_{jc} - \theta_{cs} \quad \text{(in } °C/W) \tag{19-10}$$

Table 19-1. Insulator Thermal Resistance

Insulator	θ_{cs}(°C/W)	θ_{cs}(°C/W with DC4*)
none	0.2	0.1
anodized aluminum	0.4	0.35
mica	0.8	0.4
Teflon	1.45	0.8

*DC4 is Dow Corning's silicone grease #4

20

---

reliability

RELIABILITY (*R*)

$$R = \frac{\text{Successes}}{\text{Trials}} \qquad\qquad (20\text{-}1)$$

For 67 successes in 90 trials, reliability is $67/90 = 0.744$.

MEAN TIME BEFORE FAILURE (MTBF)

$$\text{MTBF} = \sum_{1}^{n} t_f/n \qquad\qquad (20\text{-}2)$$

where
  t_f is the operational time or cycles before failure of a unit
  n is the number of units

FAILURE RATE (*f_r*)

$$f_r = \frac{1}{\text{MTBF}} \qquad\qquad (20\text{-}3)$$

RELIABILITY CALCULATION OF ONE SYSTEM

$$P_o = e^{-t/T} \qquad\qquad (20\text{-}4)$$

where
P_o = probability of success
t = mission time
T = MTBF (mean time between failures)

REDUNDANCY OF N PARALLEL SYSTEMS _____

$$P_n = 1 - (1 - P_o)^n \tag{20-5}$$

where
P_n is the probability of success of N systems in parallel
P_o is the probability of success of one system
N is the number of systems in parallel

21

---

mathematics

LOGARITHMS

$$b^x = N \qquad (21\text{-}1)$$

x is the logarithm of N to the base b and b is a finite and positive number

or

$$b = \sqrt[x]{N} \qquad (21\text{-}2)$$

EXAMPLE:

Let $b = 10$ (common logarithm) and $N = 57$.

The common log of 57 is the root x. Determine the $\log_{10}$ of a number by iterating on 10^x until suitably close to 57, as follows.

$10^{1.5}$	31.6
$10^{1.7}$	50.12
$10^{1.75}$	56.23
$10^{1.755}$	56.88
$10^{1.757}$	57.147
$10^{1.756}$	57.016
etc.	

Answer $= 1.7558749$

RULES

$$\log_b MN = \log_b M + \log_b N \qquad \text{(21-3)}$$

$$\log_b (M/N) = \log_b M - \log_b N \qquad \text{(21-4)}$$

$$\log_b N^p = p \log_b N \qquad \text{(21-5)}$$

$$\log_b b^N = N \qquad \text{(21-6)}$$

$$b^{\log_b N} = N \qquad \text{(21-7)}$$

$$\log_b \sqrt[r]{N^p} = (p/r) \log_b N \qquad \text{(21-8)}$$

$$\log_b N = (\log_a N)/(\log_a b) \qquad \text{(21-9)}$$

EXAMPLE:

The 4th root of 100 is

$$\log_{10} \sqrt[4]{100} = (\tfrac{1}{4}) \log_{10} 100 = \tfrac{1}{2}$$

anl $\tfrac{1}{2} = 3.16227$

Common or Briggsian log base is 10 (Its symbol is log *or* Log *or* $\log_{10}$). The natural, hyperbolic, or Napierian log base is *e*, or 2.7183 (Its symbol is ln).

NATURAL LOGARITHM (ln)

The natural logarithm is 2.3 times the common logarithm.

EXAMPLE:

$$\log_{10} 5 = 0.69897$$

$$\ln 5 = 1.6094379$$

$$\frac{\ln 5}{\log_{10} 5} = 2.3025851$$

THE DECIBEL (dB)

$$dB = 10 \log_{10} \frac{P_2}{P_1} \qquad \text{(21-10)}$$

$$= 20 \log_{10} \frac{V_2}{V_1} \qquad \text{(21-11)}$$

$$= 20 \log_{10} \frac{I_2}{I_1} \qquad \text{(21-12)}$$

SIGNAL LEVEL CONVERSIONS

dBM TO MICROVOLTS AND WATTS ACROSS AN IMPEDANCE Z_i

dBm is dB referred to 1 milliwatt (0.001 watts).
dBW is dB referred to 1 watt and is equal to 30 dBm.

Convert dBW to Power in Watts

$$dBw = 10 \log_{10} \frac{P}{1 \text{ watt}} \tag{21-13}$$

or

$$\frac{dBw}{10} = \log_{10} \frac{P}{1 \text{ watt}} \tag{21-14}$$

$$P = \text{power} = \log^{-1} \left(\frac{dBW}{10} \right) \tag{21-15}$$

Convert Power (P) to Voltage (e)

$$e = \sqrt{PZ_i} \tag{21-16}$$

EXAMPLE:

Convert -50 dBm to input voltage across 50 ohms.

$$-50 \text{ dBm} = -80 \text{ dBW}$$

$$-80 \text{ dBW} = 10 \log_{10} \left(\frac{P}{1 \text{ watt}} \right)$$

$$P = \log^{-1} (-8) \quad (\text{in } watts)$$

$$= 10^{-8} \text{ watts} = 10 \text{ nanowatts}$$

$$e = \sqrt{P \cdot 50} = \sqrt{10^{-8} \cdot 50} = \sqrt{5 \cdot 10^{-7}}$$

$$= 7.07 \cdot 10^{-4} \text{ volts} = 0.707 \text{ millivolts}$$

When a signal from a generator with an impedance equal to the load is used, a 3 dB loss must be included.

AREA

TRAPEZOIDAL RULE

$$A = \int_a^b f(x) \, dx$$

$$\approx \Delta x \left[\frac{f(a)}{2} + f(x_1) + f(s_2) + f(x_3) + \cdots f(x_{n-1}) + \frac{f(b)}{2} \right] \tag{21-17}$$

$$\Delta x = \frac{(b - a)}{n} \tag{21-18}$$

IF Bandpass Example ($y = f(x)$)

a and b are the starting and finishing points
n is the number of intervals (here $n = 7$)

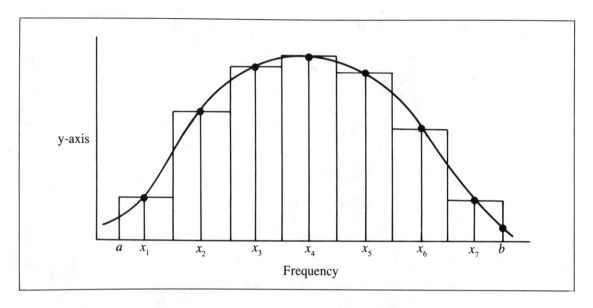

Effective Noise Bandwidth Example

The power/frequency product or area of a rectangle whose height is k and width is w is the effective noise bandwidth of the IF example previously shown.

Area = $A = kw$
Amplitude = k
Width = A/k

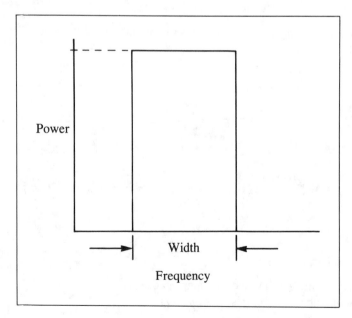

RECTANGULAR RULE

$$A = \int_a^b f(x)\, dx \tag{21-19}$$

$$\approx \Delta x \left[f(x_1) + f(x_2) + f(x_3) + \cdots f(x_n) \right]$$

Divide the area into n intervals.
x_n is the magnitude of the midpoint of each interval

$$\Delta x = \frac{b - a}{n} \tag{21-20}$$

ERROR FUNCTION

$$erf(x) = \frac{1}{\sqrt{2\pi}} \int_{-\infty}^{x} e^{-y^2/2}\, dy \tag{21-21}$$

$$erf(x) = 1 - erfc(x) \tag{21-22}$$

where
$erfc(x)$ is the complementary error function of x.

COMPLEMENTARY ERROR FUNCTION

$$erfc(x) = 1 - erf(x) \tag{21-23}$$

$$erfc(x) = \frac{1}{\sqrt{2\pi}} \int_{x}^{\alpha} e^{-y^2/2}\, dy \tag{21-24}$$

APPROXIMATION OF $erfc(x)$

$$erfc(x) = \frac{1}{x\sqrt{2\pi}} \left(1 - \frac{1}{x^2} \right) e^{-x^2/2} \quad \text{(for } x > 2) \tag{21-25}$$

For a 10-percent error, x is 2, and for a 1-percent error, x is greater than 3.

FOURIER SERIES

Conditions: The waveform must be periodic, single-valued and continuous, without an infinite number of maxima or minima near any point. A finite number of discontinuities are allowed within the waveform.

The waveform $y = f(x)$ can be represented by a series consisting of its harmonic content. For $n = 1$ to a finite value,

$$y = f(x) = A_0 + A_n \sin(nx) + B_n \cos(nx) \tag{21-26}$$

The n value represents the harmonic number
A_0 is the DC value

The value of A or B represents the magnitude of the harmonic component

$$A_0 = \frac{1}{2\pi} \int_0^{2\pi} y(dx) \tag{21-27}$$

$$A_n = \frac{1}{\pi} \int_0^{2\pi} y \sin(nx) \, dx \quad (n \geq 1) \tag{21-28}$$

$$B_n = \frac{1}{\pi} \int_0^{2\pi} y \cos(nx) \, dx \quad (n \geq 1) \tag{21-29}$$

POLES AND ZEROS

The *zeros* are the roots of the numerator of a factored polynomial (0). *Poles* are the roots of the denominator of a factored polynomial (x). Real roots can exist singly, while complex roots must be in pairs. Poles are restricted in the left-hand plane for network stability. Zeros can exist in any plane.

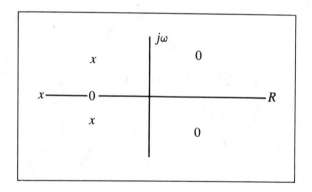

KELVIN TEMPERATURE SCALE

$$°K = °C + 273.15 \tag{21-30}$$

EXAMPLE

$$1063°C = 1336.15°K$$

or

$$90.18°K = 90.18 - 273.15$$
$$= -182.97°C$$

CELSIUS AND FAHRENHEIT

$$°C = \tfrac{5}{9}(°F - 32) \tag{21-31}$$
$$°F = \tfrac{9}{5}(°C) + 32 \tag{21-32}$$

A

greek alphabet

Lowercase	Capital	Name
α or a	A	alpha
β	B	beta
γ	Γ	gamma
δ	Δ	delta
ϵ	E	epsilon
ζ	Z	zeta
η	H	eta
θ or ϑ	Θ	theta
ι	I	iota
κ or $\varkappa$	K	kappa
λ	Λ	lambda
μ	M	mu
ν	N	nu
ξ	Ξ	xi
o	O	omicron
π	Π	pi
ρ	P	rho
σ or s	Σ	sigma
τ	T	tau
υ	Υ	upsilon
ϕ or φ	Φ	phi
χ	X	chi
ψ	Ψ	psi
ω	Ω	omega

B

---

electronics properties

Many of the more definitely assigned variables are listed in this appendix in terms of (1) the symbol used to depict the variable or property and (2) the unit in which the property is quantified followed by the unit's abbreviation in parentheses. The last (right-hand) column states the name of the property or variable.

Because of the hundreds of equations in this handbook and the number of variables in each equation, there are thousands of variables. Therefore, some variables must be reused for several different purposes. The reader is cautioned to select definitions appropriate to the application. To avoid possible conflict, refer to the definitions either directly below the equation or in other material of the same section.

In this appendix, the Greek symbols lead the list, followed by an alphabetized listing of the other variables.

GREEK SYMBOLS

Symbol	Unit	Property
α	ratio or decibels (dB)	attenuation
β	radians (rad)	angle
γ	fraction	tuning diode power law
Γ	magnitude and angle	reflection coefficient
δ	—	damping factor
η	numeric	efficiency
Δ	—	change of
ΔF	kilohertz (kHz)	change of frequency
ϵ_{eff}	numeric	effective dielectric constant

Symbol	Unit	Property
θ	volts (V)	tuning diode junction contact potential
θ	degrees (θ) or radians (rad)	angle
θ_{cs}	degrees Celsius per watt (°C/W)	transistor case-to-sink thermal resistance
θ_{ja}	degrees Celsius per watt (°C/W)	transistor junction-to-air thermal resistance
θ_{jc}	degrees Celsius per watt (°C/W)	transistor junction-to-case thermal resistance
θ_{sa}	degrees Celsius per watt (°C/W)	heatsink-to-air thermal resistance
λ	feet (ft) or meters (m)	wavelength
λ_g	feet (ft) or meters (m)	waveguide wavelength
μ	henries per meter (H/m)	permeability
π	3.1415927	pi
π	—	network configuration
ρ	microhms per centimeter cubed ($\mu\Omega/cm^3$)	resistivity
σ	area	radar cross section of the target
τ	seconds (sec)	time delay
ϕ	numeric	magnetic lines of force
ω	radians (rad)	angle, $2\pi f$
ω_n	hertz (Hz)	natural loop frequency
Ω	ohms	resistance

ROMAN SYMBOLS

Symbol	Unit	Property
A	numeric or decibel (dB)	voltage gain
A_c	square inches (in²)	core area
A_e	square centimeters (cm²)	magnetic path area
A_l	mH per 1000 turns	magnetic core induction index
A_n	numeric or decibels (dB)	gain of channel n
anl	numeric	inverse logarithm
$A_{(ol)}$	numeric or decibels (dB)	open loop gain
A_r	square centimeters (cm²) or inches (in²), etc.	area of receive antenna
ASK	title	*a*mplitude *s*hift *k*eying
A_t	square centimeters (cm²) or inches (in²), etc.	area of transmit antenna
B	hertz (Hz)	bandwidth
B	lines per square inch (lines/in²)	flux density

Symbol	Unit	Property
B	lines per square centimeter (lines/cm^2)	flux density
B_m	gausses	flux density
BT	numeric	bandwidth times pulse width
BTU	The amount of energy required to raise 1 lb. of water one degree F	British thermal unit
BW	hertz (Hz), kilohertz (kHz), megahertz (MHz), etc.	bandwidth
c	$3 \cdot 10^8$ meters per second	velocity of light
C	numeric	Linvill C factor
C	farads (F)	capacitance
°C	degrees	temperature Centigrade
C_c	farads (F)	tuning diode case capacitance
C_d	farads (F)	distributed capacitance
CFM	—	*Cubic feet per minute*
C_j	farads (F)	tuning diode voltage variable capacity
cm	centimeters	0.01 meter
CMR	numeric or decibels (dB)	*common mode rejection*
cosh	numeric	hyperbolic cosine
d	—	field-effect transistor drain
D	—	field-effect transistor drain
dB	—	decibel
dBm	—	decibels relative to one milliwatt
dBW	—	decibels relative to one watt
DC4	—	Dow Corning silicone grease no. 4
DSB	—	*double sideband*
dt	numeric	change in time
e, ϵ	2.7182818	base of natural logarithm
e_b/N_o	numeric or decibel (dB)	energy per bit to noise density
e_C	volts (V)	capacitor voltage
e	volts (V)	ac output voltage
e_L	volts (V)	ac volts across inductor
ENB	hertz (*Hz*)	effective noise bandwidth
e_o or e_{out}	volts (V)	ac output voltage
e_r	volts (V)	ac volts across resistor
erf	numeric	error function
erfc	numeric	complementary error function
ERP	watts (W) or kilowatts (kW)	*effective radiated power*

Symbol	Unit	Property
ESD	numeric or decibel (dB)	*e*quivalent *s*ignal *d*ifference
E_t	volts (V)	threshold voltage
exp (*n*)	—	$e^{(n)}$
f	hertz (Hz)	frequency
F	ampere-turns (NI)	magnetomotive force
F	numeric	noise factor
F	force	mass times acceleration
°F	degrees	degrees Fahrenheit
f_b	hertz	beat frequency
f_c	hertz	cutoff frequency
FCO	hertz	cutoff frequency
f_d	hertz	tuning-diode resonant frequency
f_d	hertz	doppler frequency
F_{IF}	hertz	intermediate frequency
f_l	hertz	local-oscillator frequency
f_m	hertz	pin-diode low-frequency operating limit
FM	hertz	frequency modulation
F_n	numeric	noise factor of stage *n*
f_o	hertz	self-resonant frequency
F_p	inches of water	fan pressure head
f_r	1/MTBF	failure rate
f_r	hertz	tunnel-diode resistive cutoff frequency
f_r	hertz	receiver frequency
f_s	hertz	spurious frequency
FSK	—	*f*requency *s*hift *k*eying
F_∞	hertz	frequency of infinite attenuation
g	32 ft/sec/sec	gravitational constant
G	—	field-effect transistor gate
G	numeric or decibels (dB)	gain
G	mhos (℧) or siemens (S)	conductance
g_m	micromhos (μ℧) or microsiemens (μS)	transconductance
g_{fx}	mhos (℧) or siemens (S)	transconductance
GHz	giga hertz (10^9 hertz)	frequency
G_{max}	numeric or decibels (dB)	maximum available gain
G_n	numeric or decibels (dB)	gain of stage *n*
G_R	decibels (dB)	receiving antenna gain
G_T	decibels (dB)	transmitting antenna gain
G_u	numeric or decibels (dB)	unilateral gain
G_v	numeric or decibels (dB)	voltage gain

Symbol	**Unit**	**Property**
G_2	decibels (dB)	gain of stage 2
H	ampere-turns per inch	magnetomotive force
H	Oersteads (gilberts per cm) (Oe)	magnetomotive force
$H(s)$	—	loop transfer function
Hz	hertz (Hz), kilohertz (kHz), megahertz (MHz), etc.	frequency
h_{11}	ohms (Ω)	input impedance
h_{21}	numeric	forward transfer current ratio
h_{22}	mhos ($\mho$) or Siemens (S)	output admittance
h_{12}	numeric	reverse transfer voltage ratio
i	amperes (A)	current
I	amperes (A)	current
I_{ib}	amperes (A)	input bias current
I_{io}	amperes (A)	input offset current
IM	decibels (dB)	*inter*modulation
I_m	decibels relative to 1 milliwatt (dBm)	mth order intercept point
I_n	amperes (A)	nth channel input current
I_p	amperes (A)	peak current
I_s	amperes (A)	saturation current
I_t^2	decibels relative to 1 milliwatt (dBm)	total 2nd-order intercept point
I_t^3	decibels relative to 1 milliwatt (dBm)	total 3rd-order intercept point
I_v	amperes (A)	valley current
I_1^2	decibels relative to 1 milliwatt (dBm)	first-stage 2nd-order intercept point
I_2^2	decibels relative to 1 milliwatt (dBm)	second-stage 2nd-order intercept point
j	-1	j operator
$J_n(\beta)$	—	Bessel function of the first kind of the nth order and β is the argument
k	numeric	coefficient of coupling
k	$1.38 \cdot 10^{-23}$ joules per degree K (J/°K)	Boltzman's constant
K	$1.38 \cdot 10^{-23}$ joules per degree K (J/°K)	Boltzman's constant

Symbol	Unit	Property
K	constant	filter section type
K	numeric	Nagaka's constant
K	numeric	Rollett's stability factor
K_{HOC}	numeric	Hall output sensitivity constant
K_o	radians per volt per second (rad/V/s)	voltage-controlled oscillator sensitivity
°K	degrees Celsius (°C) + 273.18	degrees Kelvin
K_d	volts per radian (V/rad)	phase-sensitive detector gain
l	centimeters (cm), inches (in), etc.	length
L	henries (H)	inductance
L	numeric	loss ratio
L	numeric	atmospheric loss ratio
ln	numeric	natural logarithm
$\log_{10}$	numeric	logarithm base 10
$\log^{-1}$	numeric	inverse log (or *antilog*)
L_s	henries (H)	lead inductance
m	numeric	meters
m	constant	derived filter type
m	integer	harmonic number
M	henries (H)	mutual inductance
MAG	numeric or decibels (dB)	*m*aximum *a*vailable power *g*ain
MHz	10^6 Hertz	frequency
mm	0.1 centimeter	millimeter
MNI	numeric or decibels (dB)	*m*odulation *n*oise *i*mprovement factor
MSG	numeric or decibels (dB)	*m*aximum *s*table *g*ain
MTBF	hours, days, years	*m*ean *t*ime *b*etween *f*ailure
n	numeric	number of turns
n	numeric	variable
n	numeric	turns ratio
n	integer	harmonic number
NF	decibels (dB)	*n*oise *f*igure
N_i	decibels referred to 1 milliwatt (dBm)	input noise power
N_o	decibels referred to 1 milliwatt (dBm)	output noise power
N_p	turns	primary turns
N_p	nepers (Np)	0.1151 times dB
ns	nanoseconds (ns) = 10^{-9} seconds	time

Symbol	Unit	Property
N_s	turns	secondary turns
N_{tap}	turns	tap turns
N_{total}	turns	total turns
OOK	—	*on/off keying*
P	watts (W)	instantaneous power
P	watts (W)	power
P_{avg}	watts (W)	average power
P_c/N	numeric	carrier-power-to-noise-power ratio
PCM	—	*pulse code modulation*
P_d	watts (W) per unit area	power density
P_d	watts (W)	power dissipated
P_e	—	probability of error
P_f	numeric	power factor
P_{fa}	numeric	probability of false alarm
P_g	numeric	power gain for any source any load resistance
P_n	decibels referred to 1 milliwatt (dBm) or watts (W)	noise power
P_n	numeric	probability of success of n parallel systems
P_o	numeric	probability of success
P_r	decibels referred to 1 milliwatt (dBm) or decibels referred to 1 watt (dBW) or watts (W)	power received
$P_{r\ min}$	decibels referred to 1 milliwatt (dBm) or watts (W)	minimum received power
PSK	—	*phase-shift keying*
P_t	decibels referred to 1 milliwatt (dBm) or decibels referred to 1 watt (dBW) or watts (W)	power transmitted
q	coulombs (C)	instantaneous charge
Q	numeric	quality factor
Q	numeric	junction quality factor
r	centimeters (cm), inches (in), etc.	wire radius
R	ohms (Ω)	resistance

Symbol	Unit	Property
R	rels (R)	reluctance
R	feet (ft), nautical miles (nm), etc.	range to target
R	percent (%)	reliability
r_{b1}	ohms (Ω)	base-1 resistance
r_{b2}	ohms (Ω)	base-2 resistance
r_{ds}	ohms (Ω)	drain-to-source resistance
R_e	numeric	real part of
R_e	ohms (Ω)	emitter resistance
$R_{L\,opt}$	ohms (Ω)	optimum-load resistance for match
R_{max}	feet (ft), miles (mi), etc.	maximum range
R_{ml}	magnitude and angle	conjugate match output reflection coefficient
rms	numeric	root mean square
R_{ms}	magnitude and angle	conjugate match input reflection coefficient
R_p	ohms (Ω)	tuning-diode junction parasitic resistance
R_s	ohms (Ω)	tuning diode series resistance
r_{s1}	magnitude and angle	input plane stability circle center
R_{s1}	magnitude	input plane stability circle radius
r_{s2}	magnitude and angle	output plane stability circle center
R_{s2}	magnitude	output plane stability circle radius
$R_{s\,opt}$	ohms	optimum source resistance for match
R_u	feet (ft), miles (mi), etc.	unambiguous range
RZ	—	return to zero
r_{02}	magnitude and angle	center of constant-gain circle
R_{02}	magnitude	radius of constant-gain circle
s	—	field-effect transistor source
S	—	field-effect transistor source
S	S units	signal strength
(s)	—	complex frequency variable
SFDR	decibels (dB)	spurious free dynamic range
S_i	decibels referred to 1 milliwatt (dBm) or watts (W)	signal power input
$S_{i\,min}$	decibels referred to 1 milliwatt (dBm)	minimum received signal
sinad	ratio or decibels (dB)	signal plus noise and distortion to noise and distortion
sinh	numeric	hyperbolic sine
S/N	ratio or decibels (dB)	signal-to-noise ratio

Symbol	Unit	Property
S_o	decibels referred to 1 milliwatt (dBm) or watts (W)	signal power output
SSB	—	single sideband
S_{11}	magnitude and angle	input reflection coefficient (3-port, ports 2 and 3 matched)
S_{11}	magnitude and angle	input reflection coefficient (2-port, output matched)
S_{12}	magnitude and angle	reverse transfer coefficient (output matched)
S_{21}	magnitude and angle	forward transfer coefficient (ports 1 to 2, 3-port, ports 2 and 3 matched)
S_{21}	magnitude and angle	forward transfer coefficient (output matched)
S_{22}	magnitude and angle	output reflection coefficient (input matched)
S_{31}	magnitude and angle	forward transfer coefficient (ports 1 to 3, 3-port, ports 2 and 3 matched)
SW	—	switch
t	degrees	temperature
T	seconds (s)	time constant
T	seconds (s)	pulse width
T	—	network configuration
T_a	degrees Kelvin (°K)	ambient temperature
tanh	numeric	hyperbolic tangent
T_d	seconds (s)	time delay
T_e	degrees kelvin (°K)	effective noise temperature
TE	—	*t*ransverse *e*lectric mode
T_{fa}	seconds (s)	time between false alarms
T_j	degrees	junction temperature
TM	—	*t*ransverse *m*agnetic mode
T_r	degrees	air stream temperature rise
v	volts (V)	instaneous voltage
V	volts (V)	voltage
VA	—	volt-amperes
V_{BB}	volts (V)	unijunction transistor supply voltage
V_{CC}	volts (V)	power supply voltage
VCO	—	*v*oltage-*c*ontrolled *o*scillator
V_D	volts (V)	diode contact potential
V_f	volts (V)	tuning voltage
V_H	volts (V)	Hall output voltage

Symbol	Unit	Property
V_{io}	volts (V)	operational amplifier input offset voltage
V_p	volts (V)	peak voltage
V_p	percent (%)	velocity of propagation
V_{pk}	volts (V)	peak ac volts
VSWR	ratio	*voltage standing wave ratio*
VU	—	*volume unit*
W	ergs (e) or joules (J)	power converted to heat
X_C	ohms	capacitive reactance
X_L	ohms	inductive reactance
X_n	ohms	nth reactance
X_t	ohms	total reactance
y	mhos (v)	admittance
y_{fb}	mhos (v)	common-base forward transfer admittance
y_{fc}	mhos (v)	common-collector forward transfer admittance
y_{ib}	mhos (v)	common-base input admittance
y_{ic}	mhos (v)	common-collector input admittance
y_{ob}	mhos (v)	common-base output admittance
y_{oc}	mhos (v)	common-collector output admittance
y_{rb}	mhos (v)	common-base reverse transfer admittance
y_{rc}	mhos (v)	common-collector reverse transfer admittance
y_{11}	mhos (v)	input admittance
y_{12}	mhos (v)	reverse-transfer admittance
y_{21}	mhos (v)	forward-transfer admittance
y_{22}	mhos (v)	output admittance
Z	ohms (Ω)	impedance
Z_{in}	ohms (Ω)	input impedance
$Z_{i(ol)}$	ohms (Ω)	open-loop input impedance
Z_n	ohms (Ω)	channel n input impedance
$Z_{o(ol)}$	ohms (Ω)	open-loop output impedance
Z_{out}	ohms (Ω)	output impedance
z_{11}	ohms (Ω)	input impedance
z_{12}	ohms (Ω)	reverse transfer impedance
z_{21}	ohms (Ω)	forward transfer impedance
z_{22}	ohms (Ω)	output impedance

bibliography

Atkins, Bob. "A Practical Dish Feed for the Higher Microwave Bands." *QST* (February 1981): 63.

Bahl, Dr. I.J. and D.K. Trivedi. "A Designers Guide to Microstrip Line." *Microwaves* (May 1977): 174-82.

Bell, F.W., Inc. *The Hall Effect and its Applications*. Columbus, Ohio: Bell, Inc. Publication.

Buchbaum, Walter H., Sc.D. *Buchbaum's Complete Handbook of Practical Electronic Reference Data*. Englewood Cliffs, New Jersey: Prentice Hall, Inc., 1973.

Carlson, A. Bruce. *Communication Systems*. New York: McGraw-Hill, 1968.

Chauvin, Ronald. "Find Pulse Droop Fast." *The Electronic Engineer* (September 1986): 60-1.

Davis, Frank. "Matching Network Designs With Computer Solutions." *Motorla Application Note AN-267*.

Doyle, Norman P. "Swift Sure Design of Active Bandpass Filters." *EDN* (January 15, 1970): 43-50.

Dumas, Kenneth L. and Leo G. Sands. *Microwave Systems Planning*. New York: Hayden Book Company, Inc., 1967.

EDN Editorial Staff. "Heat Loss by Radiation." *EDN* (November 25, 1968): 62.

Fisk, James. "Simple Formula for Microstrip Impedance." *Ham Radio* (December 1977): 72-3.

Froehner, William. "Quick Amplifier Design with Scattering Parameters." *Electronics* (October 16, 1967): 5-2 to 5-11.

Hatchett, John. "Calculate Capacitor Tap Impedance with Correct Expression and Avoid Errors." *Electronic Design* 8 (April 12, 1975): 82.

Hewlett Packard Staff. "RF and Microwave Diode Applications Seminar." *Hewlett Packard Corporation.*

Hewlett Packard Staff. "The Hot Carrier Diode—Theory, Design, and Application." *Hewlett Packard Application Note 907.*

Hewlett Packard Staff. "S Parameter Circuit Analysis and Design." *Hewlett Packard Corporation* (September 1968): 1-1 to 7-11.

Kuecken John A. *Antenna and Transmission Lines.* Indianapolis, Indiana: Howard W. Sams and Co., Inc., 1969.

Lange, Julius. "Microwave Transistor Characterization Including S Parameters." *Texas Instruments, Inc. under U.S. Army Electronics Command, Fort Monmouth, New Jersey, Contract DA28-043 AMC-01371(E).*

Magnetics, Inc. Staff. "Designing DC-DC Converters." *Magnetics Inc. Publication.*

Manken Arthur H. "Selecting Air Movers Without Guess Work." *Electronic Design* 11 (May 24, 1969): 90-3.

McVay, Franz C. "Don't Guess the Spurious Level." *Electronic Design* 3 (February 1967): 70-3.

Mitchell D.C. "HP-67/97 Tracks Communication Satellites." *Electronics* (March 1, 1979): 146.

Motorla Application Engineering. "Epicap Tuning Diode Theory and Application." *Motorola Semiconductor Products, Inc. Technical Information Note AN-178A.*

Norton, Dr. David E. "The Cascading of High Dynamic Range Amplifiers." *Adams Russel Electronics Co. Inc., Anzac Electronics Division Catalog Application Note* (1989): 14-15.

Pamotor Staff. "Determining the Airflow and Pressure Requirements." *Pamotor Burlingame California Bulletin 7041* (1970): 3-10.

Rao, M.V. Subba. "Circuit Converts Unipolar Digital Data to Alternate Mark Inverse Format." *Electronic Design* 4 (February 15, 1974): 106.

Schulz, Walter J., Jr. "HP25/HP-33E Aids Design of Short Vertical Antennas." *Electronics* (February 1, 1979): 135–7.

Schwartz, Mischa. *Information Transmission Modulation and Noise*. New York: McGraw-Hill, 1970.

Skolnik, Merrill I. *Introduction to Radar Systems*. New York: McGraw-Hill, 1962.

Spilker James J. *Digital Communications by Satellite*. Englewood Cliffs, New Jersey: Prentice Hall Inc., 1977.

Teledyne Semiconductor Staff. "FET Small Signal Analysis—J FET Applications and Specifications." *Teledyne Semiconductors Mountain View, California* (June 1972): 110-3.

Tranbarger, Oren. "Symmetrical Bridge Subs as Balun." *EDN* (October 1967): 68-71.

Westman, H.P., Editor. *Reference Data for Radio Engineers*. Indianapolis, Indiana: Howard W. Sams and Co., Inc., 1969.

Wilson, Robert C. K7ISA and Hal Silverman W3HWC. "Wire Line—A New Easy Method of Microwave Circuit Construction." *QST* (July 1981): 21-3.

index _____